# The Terror of Tobermory

An informal biography of Vice-Admiral
*Sir Gilbert Stephenson, KBE, CB, CMG*

## Richard Baker

**BIRLINN**

This edition first published in 1999 by
Birlinn Limited
West Newington House
10 Newington Road
Edinburgh EH9 1QS

Reprinted 2002, 2015

First published in 1972 by W.H. Allen

ISBN 9781780273020

British Library Cataloguing-in-Publication Data
A catalogue record for this book is available
from the British Library

# Contents

# Author's Preface

Not long ago, in a hotel foyer, I caught sight of a lady of mature years pointing me out to a small boy. The boy came over to me and said politely, 'Is your name Richard Baker?' I replied that it was, whereupon the boy said, 'My Gran remembers you reading the news. She says you're historical.'

Well, facts must be faced. It's true that I have lived through a fair amount of history. Old folks are notorious for remembering the events of their youth far more clearly than what happened yesterday, and I recall quite vividly a number of visits at the nation's expense to Scotland in World War Two.

As an Ordinary Seaman in the Navy, I joined a fleet minesweeper at Granton, on the Firth of Forth; several years later as a Sub-Lieutenant, I was appointed Entertainments Officer at the large naval air station at Macrihanish on the Mull of Kintyre, after which I crossed the country to serve as Educational and Vocational Training officer at another naval air station at Rattray in Aberdeenshire. In between, as a Midshipman, I served in the sloop HMS *Peacock*. From our base at Greenock, on the Clyde, we escorted several transatlantic convoys, and in the summer of 1944 we set out from Loch Ewe as part of the escorting force for Convoy JW59, bound for north Russia.

Our ship had been prepared for its operational role with a brief but arduous spell of training supervised by the subject of this book from his headquarters ship, HMS *Western Isles*, moored in Tobermory Bay on the Isle of Mull. Vice-Admiral Sir Gilbert Stephenson, who ruled the roost in those days as Commodore, *Western Isles*, was a truly remarkable character. He terrified the life out of many a matelot and naval officer, but when I told him, 25 years after the end of the war, that I would like to write a book inspired by his legendary ferocity, he assumed a look of shocked surprise and replied with a twinkle of piercing blue eyes: 'But I am the mildest man you could ever hope to meet!'

Tobermory is a quiet, attractive small town with houses strung out along the edge of a lovely, almost landlocked bay. But between 1940 and 1945 that bay was alive with incessant naval activity. About a thousand small ships – destroyers, sloops, frigates, corvettes, minesweepers and the like – came there for a fortnight's intensive 'work-up' before going out to face the U-Boat war in the Atlantic and Western Approaches.

For many of the men in those ships, their trip to Tobermory was literally the first time they had been to sea. In fourteen days they had to learn the essentials of seamanship, to weld themselves into a team, and to master the complicated techniques of naval warfare. The man who had to make sure this impossible task was achieved worked like a demon and drove everyone else to do likewise. Woe betide anyone who did not react quickly enough to an order: the great voice would ring out – 'When I say move, I mean MOVE!!!' At this period of his life, Sir Gilbert Stephenson inspired many nicknames, not all as polite as 'The Terror of Tobermory'.

Nor was this force of character felt only by the students in Tobermory Bay. The thunder rolled more than once along the corridors of the Admiralty. For if Their Lordships disagreed with the policy of the base, its Commanding Officer would 'proceed independently' – a vigorous trait which could be traced back right through Sir Gilbert's career, and probably cost him the highest promotion. But as Sir Gilbert said: 'I never wanted to be Admiralissimo. I had no ambitions except perhaps to do the job I was put to well.'

Of course it was essential to Sir Gilbert's purpose at Tobermory to be reckoned a Terror by his victims. He gave a very well studied blood-and-thunder performance designed to drive the ABC of seagoing efficiency into people like myself, who had barely the first notion of naval ways. My own first encounter with him is still a vivid memory.

In the Wardroom of HMS *Peacock*, a sloop of the Black Swan class, we had sought relief from the ardours of our 'work-up' in a rather wild party, in the course of which I had been copiously lathered with gentian violet ointment. All visible parts of me were bright mauve, and remained so for about a week. My Captain had a perverse sense of humour, and decided to send me in this condition to deliver a message to the Commodore on board his flagship.

It was humiliating enough to be transported across the bay by a boat's crew of scornful sailors, worse to board HMS *Western Isles* and be saluted uncertainly by a transfixed quartermaster; the thought of facing the Commodore was almost unbearable. However it had to be done.

I was ushered into his cabin, delivered my message, and was about to depart as briskly as I could when the Commodore halted me.

'What's your name, sonny?'

I told him.

'And what's your job on board?'

'I'm Assistant Gunnery Control Officer, sir, and Entertainments Officer.'

'Oh I see,' said the Commodore (and was there a trace of a smile on that eagle-eyed countenance?), 'Well, it's clear you take the second part of your job very seriously!'

I thought I was lucky to escape with my life.

Long before a ship reached Tobermory, the buzz would go round about the terrifying old gentleman (Sir Gilbert was well into his sixties) who was in command there. It was known that he made a habit of descending on ships in harbour quite unannounced, and would create hell if the Officer of the Day was not there to meet him. Well-briefed ships took very good care to keep a sharp lookout for the blue and white barge that flew the pennant of 'Commodore, *Western Isles*'.

Many stories are told about the Commodore's sudden visits – I have included some of them later in this book. The best known has been elaborated in many different ways, but this is the Authorised Version:

One winter morning, the lookout in a corvette gave warning of the approach of the Commodore's barge – in sufficient time for the Captain as well as the Officer of the Day and the Quartermaster to be lined up smartly on deck when the Commodore came on board.

Without any preliminaries, he flung his goldbraided cap on the deck and said abruptly to the Quartermaster: 'That is a small unexploded bomb dropped by an enemy plane. What are you going to do about it?'

The sailor, who had evidently heard about these unconventional tests of initiative, promptly took a step forward and kicked the cap into the sea.

Everyone expected a great roar of protest from the Commodore. But not at all. He warmly commended the lad's presence of mind, and then, pointing to the rapidly sinking cap, said: 'That's a man overboard. What are you going to do about that, eh?'

It was November in the north of Scotland, but the boy had to dive in and rescue that cap.

Sir Gilbert's approach to the problem of turning inexperienced men into fighting sailors in fourteen days was an undoubted success. His knighthood in 1943 was bestowed in recognition of the contribution he made to victory in the Battle of the Atlantic, and his working-up methods influenced subsequent procedures in the Royal Navy.

I hope many people of the present generation will find it interesting to read about someone who symbolised a Navy and a Britain which have vanished for ever. He himself was very good at bridging the generation gap. He was endowed with a challenging mind, which broke down the barriers which normally divide ninety-odd from nineteen, and anyone with half-digested assumptions about life was likely to be quizzed without mercy.

This is not a formal biography. It is simply an attempt to capture the personal quality of one of the great characters of his time. Imagine him in his study at home in Saffron Walden, Essex, a comfortable cabin of a room where we talked together over a period of about eighteen months. Every now and then, when deeply engaged, he would rise and pace the deck, a tough, upright, not very tall figure. His brilliant blue eyes flashed often with amusement or anger, the fine voice rising to an emphatic roar or shaken with tremendous laughter. Sir Gilbert received me always with great good humour (what happened to that ferocious Commodore?) and the hospitality offered by him and his daughter Nancy was a joy.

This book was originally published just a few weeks before the Admiral's death at the age of ninety-four, on May 27 1972, and I am delighted that Birlinn Ltd decided to reissue it as a paperback. I hope that in this way a new generation will learn something of the significant role played by a small Hebridean town in the greatest battle of World War Two, and will enjoy meeting that man who chose to be known as 'The Terror of Tobermory'.

Richard Baker
January 2002

# Foreword

The Royal Navy has always had 'characters' who have made its history more colourful. In the first half of the twentieth century Vice-Admiral Sir Gilbert Stephenson must definitely be included in their forefront.

'Puggy' (or 'Monkey') Stephenson always made an impact on all who served under him, and I reeled under that impact when he took over command of the *Revenge* in 1923. I had joined her in January that year when she was with the International Fleet at Constantinople. I found her a perfectly happy and efficient ship under a popular Captain.

When we left the Mediterranean in the early summer we wanted no change and least of all a new Captain, who already half a century ago had achieved the reputation of being a 'Terror'.

He bristled and yapped like a ferocious little fox terrier whose bite was soon proved to be as bad as his bark. I note that in Chapter Nine the author has quoted from a letter of mine which gives the impression that after his first inspiring speech the ship was transformed. But this transformation took time and was not without some pain under our new Captain's assaults.

My wife got to know him quicker than I did. One day she said 'Why are you all so terrified of the Captain? He is such a nice little man really and puts up a facade of temper and shouting to try and conceal a gentle, rather timid character which he feels is inappropriate to the Captain of a battleship.' In a flash I saw she was right and with this new understanding I began to get on famously with him. And it wasn't long before the whole ship's company responded and the *Revenge* rose to new heights.

Reading Chapter Nine I have discovered for the first time why all the bad hats who joined the *Revenge* were always drafted to my division when possible, and now after 48 years I appreciate the compliment. Puggy Stephenson also tells the tale of trying to get me thrown out of the ship before he took over command. But what he does not mention is that when saying 'goodbye' to me he 'confessed' to what he had tried to do. This was an uncalled for but typically generous act on his part, for I doubt if I should even have appreciated the attitude which Senior Naval officers could take to me on principle. In fact in my very next job at sea, after specialising in Communications, the Commander-in-Chief of the Mediterranean Fleet at first refused to have me on his staff, even in the humble capacity of Assistant Fleet Wireless Officer, but relented before it was too late.

However, Puggy Stephenson's career as Commodore *Western Isles* is where he earned his nickname 'The Terror of Tobermory' and where he made such a vital impact on the Battle of the Atlantic in 'working-up' green ships' companies of escort and anti-submarine vessels. He went there in 1940 and by 1941 his fame – or possibly notoriety? – became a byeword in the Fleet, where I was then serving in command of the Fifth Destroyer Flotilla. In 1942, when I was Chief of Combined Operations, the Commandos crossed swords with him. The account given in Chapter Seventeen does not include a certain incident, which they reported to me. Knowing that the Commodore made a habit of boarding ships at night who kept a slack lookout, and removing life buoys and other souvenirs which had to be reclaimed in the morning, the Commandos decided to test the vigilance on board his own flagship. It was not without some malicious amusement that I received a report that they had succeeded in removing the personal barge of the Commodore from alongside HMS *Western Isles.*

Stories continued to circulate about him and even reached me in my new Headquarters in South-East Asia.

What cannot be overstressed is the almost miraculous results he achieved in working up newly commissioned ships so quickly. One thousand, one hundred and thirty-two uncoordinated groups of mainly inexperienced landlubbers passed through his hands and in the incredibly short time of two weeks were welded into a disciplined coordinated ship's company who were good enough to go straight out to take their part in the Battle of the Atlantic. Some 130 enemy U-boats were sunk (91 confirmed) by Tobermory trained ships and they accounted for some 40 enemy aircraft. But above all a steady stream of efficiently worked-up escorts continued to be thrown, in the shortest possible time, into the absolutely vital Battle of the Atlantic, whose loss would have meant the loss of the war for us.

So may I end with the well-known Naval signal:

'Well done, Stephenson.'

*Mountbatten of Burma*

A. F.

ADMIRAL OF THE FLEET
THE EARL MOUNTBATTEN OF BURMA

# Illustrations

Seven Ages of Stephenson

*c.* 1888: Schoolboy Stephenson – already a sailor!

1897, aged 19: Kitted out for the Benin Expedition.

1916, aged 38: Commander, Crete Patrols.

1924: Captain Stephenson.

1927–8: Commodore RN Barracks, Portsmouth.

*c.* 1942: Commodore Stephenson, RNR.

November 1969: Aged 91.

HMS *Active*.

Crest from the German Vice-Consulate at Chios.

HMS *Dauntless* and HMS *Revenge*.

1916: In the Kaiser's Navy.

*A Family at War*, March 1940.

In Tobermory: HMS *Western Isles* on Christmas Day.

Tobermory.

March 17, 1944: Some of the corvettes and frigates in Tobermory Bay.

1943: Commodore Western Isles with his colleagues.

The *Western Isles* crest.

The 'Terror'.

1956: The Price of Peace.

1955: The Hon. Commodore Sea Cadet Corps is received on board.

1963: With Sir Horace Law.

HRH Prince Philip with Sir Gilbert Stephenson and Chairman-Captain R.W. Ravenhill.

1968: HMS *Eaglet*, Liverpool.

1960s: Dining with the Croydon Unit of the RNVSR.

1970: Annual Dinner of the RN and RNVR at Chatham Barracks.

May 1971: Sir Gilbert Stephenson at the Battle of the Atlantic Memorial Service in Liverpool Cathedral.

1970: Sir Gilbert Stephenson appears on TV.

Liverpool, summer 1971, aged 93: Sir Gilbert Stephenson cuts the tape.

Sir Gilbert Stephenson and the author.

Sir Gilbert Stephenson and his daughter Nancy at home.

'To tell the *truth*? Do you mean to say you don't always tell the truth?'

# PART I

## Stephenson, RN
### (1878–1928)

# 1

# A Nervous and
# Timid Person

(1878–1892)

The man who was to play such a significant part in the war
against Hitler's U-boats was born long before submarines
went into service with the world's navies. 'I am told', said Sir
Gilbert, 'that I was born in 1878; and being a simple man, I
believe it!' Disraeli was Prime Minister, and Great Britain was
confidently colouring the map red in the name of God and
Queen Victoria. The Royal Navy had been supreme at sea since
Napoleonic times.

However, Sir Gilbert's family had no naval connections. His
father did well as a merchant in India and as a company
secretary, but died quite young; so he was brought up by his
mother and his grandfather, Sir Macdonald Stephenson, who
had one of those big London houses in Lancaster Gate which
have long since been broken up into flats and bed-sitters.

They were a family with strong Christian beliefs, and the code
of duty and honour shared by so many prosperous Victorian
households. The young Gilbert was brought up with the idea
that obedience, good manners and respect for elders were as
much a part of nature as the fact that night followed day – it was
part of the order of life. Against this firm background, they were
a happy family, and very fond of one another.

Gilbert had three brothers, and the four of them got along
pretty well – though they were glad they didn't have a sister –
in case she turned out like them. One of the boys became a

1

solicitor, another a chartered accountant who later in life did a great deal of work for ex-prisoners – and the eldest went to the Bank of England – 'having' – according to the Admiral – 'even less brains than I have'.

It is clear that the great influence on young Gilbert Stephenson was his mother: 'She was a wonderful woman – very beautiful, and a marvellous pianist. No one played a more important part in my life. And one of the most wonderful things about her, after we'd been so close, was the way she turned me over, as it were, to her daughter-in-law when I got married. It can't have been easy to let me go like this. A lot of mothers can't do it, you know! I admired my mother so much for her wisdom. I was hardly aware of the pressure she exerted on me till afterwards – it was so subtle, so delicate. She spoke French very well and passed her knowledge on to us; as a matter of fact, one of my fewer clear memories of early schooldays was coming home and telling my mother that the French master knew less French than I did.'

There was little of the ferocity of the future Commodore *Western Isles* about the young Gilbert as a schoolboy. He was a pretty good left-hand bowler, but for the rest 'just slightly below average' – not rebellious or naughty, in fact very quiet, 'a nervous and timid person'.

Perhaps it was as an escape from what he felt to be his own shy nature that he took such pleasure in some of Henty's swashbuckling adventure stories, and above all in W.H.G. Kingston's series of books about naval life, in which the boy could identify with the *Three Midshipmen* who became *Three Lieutenants* and ultimately *Three Commanders*. These books made him absolutely determined to get into the Navy, though none of his relations had had anything to do with it. At the age of nine the great decision was taken – a decision the Admiral never regretted.

The 1880s were in fact a time of considerable crisis in the Royal Navy: the French had actually reached parity with the British Fleet – a very alarming state of affairs to those who believed in Pitt's dictum that the Royal Navy should be a match for any *two* potential enemies in alliance. Parliament reacted with the Naval Defence Act of 1889, which produced a programme of eight first-class battleships, two second-class battleships, nine large and twenty-nine small cruisers, four gunboats and eighteen torpedo boats – all for £21,500,000. And in 1890, the American Captain A.T. Mahan produced his book, *The Influence of Sea Power upon History* which did much to revive interest in the Royal Navy and led to the formation of the Navy League – of which Sir Gilbert Stephenson became General Secretary in 1932.

Forty years earlier, in 1892, the sea-struck lad had to sit the entrance examination for the cadets' training ship HMS *Britannia* at Dartmouth – and failed. It was a big term – almost seventy were admitted – but the young Gilbert failed: 'that shows you how brainless I am. It was to be the Law for me!'

But luck was on Gilbert's side – one or two youngsters did not take up their appointments. The door opened just a crack, and the 'nervous and timid person' managed to 'creep through'.

# 2

# Not of the Least
# Importance

(Cadet and Midshipman, 1892–1898)

In 1892, the Royal Naval College had not been built at Dartmouth, and the cadets had to live and work in two old line-of-battle ships from the days of sail, *Britannia* and *Hindustan*, moored head and stern in Dartmouth Harbour. Conditions on board were terribly unhealthy; the lads slept in hammocks jammed right up against the deck overhead with the portholes well below them. After they were called in the morning, the boys all had to take a plunge into a large bath and they were always banging into each other because their eyes were all gummed up. The sick bays were invariably full – mainly through lack of ventilation.

As for their education, it left a lot to be desired. They were taught a good deal of mathematics – for navigation – and a little engineering. There was drawing and there was seamanship but even that wasn't much. Such boatwork as there was, they had to teach themselves.

There was also a good deal more bullying than was healthy. The senior terms were allowed to bully the more junior ones, and a few people seemed to get a lot of pleasure out of ordering the younger boys to hold out their hands and then swiping them with a lanyard. Gilbert Stephenson, who hated the idea of hurting another person, did not take part.

In the Navy in those days, corporal punishment was common. One lively lad tried to duck Church Parade by telling the

Commander he was a 'Bush Baptist'! Unfortunately the Commander checked this unlikely statement with the boy's father-resulting in a dozen from the SubLieutenant. But for all that, there was probably little wilful brutality.

Gilbert did his share of fighting in *Britannia*, like the rest, and was punished for it – but in his view in a sensible way: he was made to undergo compulsory boxing instruction. This he found a useful disciplinary trick himself in later years, and he showed considerable talent for the sport. He boxed in the finals in *Britannia* – at ten stone – and a few years later was due to box for the Navy in the inter-Service finals; illness unfortunately ruled him out. But boxing was always important in his career; he ran boxing clubs in most of his ships, and in boys' clubs too.

What mattered most to the young lads in *Britannia* was not the smattering of academic education they picked up, but going to sea for the first time as a young Cadet or Midshipman. They were not of the least importance; they were looked down upon by everyone – and might have to cope, though little more than schoolboys, with boatloads of drunken libertymen. That way they learnt fast, or went under.

Gilbert Stephenson joined his first sea-going ship, HMS *Endymion*, in 1894, when he was sixteen. She was a first-class cruiser of 10,000 tons or thereabouts; and there was a story about her Captain which everyone lapped up with great gusto.

He was known as 'Billy' Barlow; he fancied himself as a boxer, and used to keep in trim by taking on his Chief Steward, who of course simply *had* to obey his Captain's summons to a contest. Now 'Billy' was much better at boxing than the Chief Steward – certainly it was he who derived the greater pleasure from these encounters, and the Chief Steward longed for revenge.

His chance came one day when 'Billy' came back on board after lunching rather too well and fell asleep in his cabin. When he woke up, he discovered it was 9pm, and decided it was high time for his evening meal. So he rang for the Chief Steward, who immediately appeared. 'Dinner, please,' ordered the Captain. At last it was he who was caught off guard when the Steward replied with mock incredulity: 'What, *again,* sir?'

*Endymion* went on a visit to Kiel, for the opening of the Kiel Canal – long before anyone dreamt there could be any trouble with Germany; and she took part in a review at Spithead which brought the British and French fleets together – though there was no love lost between them at the time. During the review there was a social disaster on board *Endymion* which made a great impression on young Stephenson.

A Midshipman had been sent with the picket-boat to bring off the guests for a party from Clarence Pier. When he had embarked all the guests he could find, the Midshipman decided there were too few of them to ensure a good evening; so he forthwith cleared Clarence Pier of every female in sight.

Now Clarence Pier in those days was a notorious haunt of the demi-monde; so the 'snotty' returned to *Endymion* with an extraordinarily ill-assorted party of respectable wives, cousins and aunts on the one hand, and ladies of doubtful virtue on the other. The result was social catastrophe. Each of the officers thought one of the others was deliberately insulting *his* guests. They wouldn't speak to each other for weeks afterwards.

The most useful experience Stephenson had at that time was as Midshipman in charge of a cutter – a great big twelve-oared boat. They took a lot of pulling, and were a tremendous training in muscle power and teamwork, as many a naval man will be only too aware.

Most of Stephenson's crew were heavily bearded and enormously tough; and of course the Coxswain knew far more about boats than he did: 'When I took charge for the first time, I could almost *feel* them thinking: "Now what sort of a little so-and-so is this? Is he going to do the job properly, or make a mess of it and let us down in front of our shipmates?" They all did their best to get to windward of us wretched little snotties, and it was up to you to prove they couldn't go too far. They were tremendously loyal, though. Once they realised you were doing your best, they would stick to you through thick and thin, and woe betide any other boat's crew who made fun of *their* Midshipman. This was my first real experience of a good relationship between officers and men – and how important that is! On the whole, I am happy to say things have worked this way throughout my career, and I put it down to a great extent to that early boatwork.'

In the Fleet at that time, there was tremendous competition

between the different boats of the same ship, and boats of different ships. No one felt the time spent painting them up and racing them was wasted – it all helped to make the crews more efficient in a real emergency. A little later, Gilbert Stephenson had to take a load of officers out to a ship off Bermuda, with a very heavy sea running – and if they hadn't all known to a man *exactly* what they were doing, he was convinced the boat would have been lost.

In 1895 Midshipman Stephenson joined HMS *Forte*, a light cruiser of the Mediterranean Fleet, based at Malta. There he learnt another lesson.

In those days, Canteen Managers were powerful people in ships – in charge of stores, and in a position to do a fair amount of fiddling. So 'considerations' of various kinds did sometimes change hands. But Stephenson was very shocked one day, on the bridge of *Forte*, when the Yeoman of Signals told him that the First Lieutenant had a house ashore, and, (with a wink) 'of course he paid no rent for it'. 'Now *I knew* for a fact that this First Lieutenant of ours was a splendid fellow and certainly would not be involved in anything of that kind. But I suddenly realised that it is simply not enough, when you are in command of men, to *be* honest – you've got to be *seen* to be honest; and later on, when I was a First Lieutenant and the canteen people used to send me up bowls of fruit from time to time, I would never accept any of them.'

Young Stephenson also had a spell as Assistant Navigator in *Forte* – and one of his tasks in this connection is a reminder of how far the Navy has travelled to the computers and guided missiles of today. As the ship entered or left harbour, the guns' crews used to practise range-estimation on prominent objects ashore; Stephenson had to check the distances on the navigational chart and dash down to the guns with the proper figures to check against their calculations. They had no mechanical or electronic aids to gunnery in those days – it was all done by guesswork.

On the subject of gunnery, visitors to Admiral Stephenson's house were sometimes surprised to learn that he always kept a loaded '45 revolver by his bed – it was attached to the bedpost by a lanyard. Once, the police got to hear about it and there was a court case – they didn't seem to think it was very safe for him to have a loaded pistol in the house. But as the Admiral said: 'What's the use of having it unloaded? I'm an old man and quite defenceless. If someone broke in – and it's perfectly possible, isn't it? – there wouldn't be much time to load, would there? And as for the danger of someone else getting hold of it – as my barrister pointed out, I wouldn't give it to Jesus Christ!' The Admiral was allowed to keep his revolver.

It was given to him by a devoted aunt in 1897, through circumstances which tell us much about the days of Imperial Britain.

While Midshipman Stephenson was serving in HMS *Forte* on the Mediterranean station, his mother fell seriously ill, and he was given leave to go home and visit her. Happily her condition soon improved, and he began to think about rejoining his ship. He had fixed a passage in a naval vessel, when he saw in the paper that *Forte* had been ordered to the Nigerian coast to take part in the Benin Expedition.

Benin was then a small West African kingdom of quite exceptional brutality, relying on the slave trade for its income, and practising a religion which demanded human sacrifice. The king had been keeping up the supply of victims by capturing young people from the surrounding countryside; some of them were subjected to a revolting form of crucifixion.

Queen Victoria, who of course at that time ruled a great part of Africa, sent a delegation to stop the king from these hideous practices: his answer was to execute all but two of the envoys, making a punitive expedition of some kind inevitable. But there was a special problem about this project: almost every European to visit those parts went down after a week or two with West Coast fever; so the proposed expedition would either have to be very long – to give people the chance to have the fever and get over it before the attack on Benin – or very short, with the idea of making an assault and getting away before the onset of the disease. The short expedition clearly had the advantage in terms of surprise; but the Army did not feel an attack could be prepared with sufficient rapidity. So the job had to be done by the Navy, and Admiral Sir Harry Rawson was ordered to take command.

All this promised *action*, and the Midshipman was absolutely determined not to miss the expedition if he could possibly avoid it. But there he was in London, with no suitable kit – and no apparent means of rejoining his ship in time.

So first of all he quickly designed himself a khaki uniform and managed to persuade Hope Brothers to run it up; he talked his kindly aunt into giving him that '45 pistol; and he managed to get a passage in a hospital ship, the *Malacca*, which was fitting out at Tilbury. He felt he had very little hope of getting to *Forte* before the action, but at any rate he was determined to try.

The voyage out was very pleasant. There was a party of young nurses on board, and two of them were particularly charming. Gilbert's interest in them led him to discover an infallible, if violent, cure for seasickness.

Almost as soon as the ship left the Thames, one of the girls began to feel ill, and was in her bunk for several days. On the fourth day she was taken on deck and laid down on a deck chair under some rugs, feeling very sorry for herself. 'At that time', the Admiral recalls, 'I was chasing the other girl round and round the upper deck. This eventually winded me, and I flung myself down with some force on what I thought was an empty deck chair covered in blankets. To my great surprise, there was a tremendous yell of fury and indignation from under the blankets, and the sick girl leapt to her feet. She was totally cured. From that moment on, she never felt the slightest twinge of seasickness!'

Thenceforward the time passed pleasantly enough, and to his great delight Midshipman Stephenson rejoined HMS *Forte* with half an hour to spare before the landing party went ashore.

During the passage up the Benin river in a stern-wheeler, he got the engineer to put an edge on his cutlass, and when the party landed at Alogbo, he was ordered to bring up the rear as they marched single file through the jungle paths. It was a distinct disadvantage in this unhealthy situation that the doctor had somehow lost his supplies. But Stephenson had with him a box of Beecham's pills and some chlorodyne, and this had to suffice the whole party for medical treatment. If one of the bearers complained of feeling unwell, the doctor would judge from the state of his eyes and his stomach whether he *needed* medication. If he did, he would get one – or both – of the

Midshipman's panaceas. If not, the treatment was even simpler: the would-be patient was given half a dozen lashes administered by his friends. In this elementary way, it seems, the health of the party was well maintained.

As the expedition progressed, so Stephenson's admiration for Sir Harry Rawson grew. It was tough going, and in those days a man in his position had every right to be *carried* through the bush, particularly as he was quite elderly and rather stout. But he insisted on walking every step of the way like the rest of the party, and that made a great impression on everyone, not least one young 'snotty'.

When the leading party neared Benin and were about to emerge from the cover of the bush, Rawson fired a war rocket. This had the intended effect of terrifying the superstitious inhabitants, who presumably attributed such a noise to the devil; many of the people fled forthwith, thus greatly reducing the bloodshed that occurred. By the time Stephenson's company reached the city it was already in British hands, deserted by the natives and largely burnt. Everywhere there were human remains – pits full of bodies in all stages of decomposition, and the stench of human blood hung in the air. Stephenson came across what had clearly been a sacrificial altar, and from it he picked up a knife that must have been used in the murderous orgies.

'My son', said Sir Gilbert, 'was in the Colonial Service and spent some years working in Nigeria; and while he was there met the grandson of that evil ruler of Benin. He told him how I had been involved in this expedition against his grandfather, thinking he would feel aggrieved by what, after all, was an exercise of imperial power on Britain's part. But the grandson's verdict was that we had been right.

'So this speedy assault on Benin was a complete success, thanks to the moral courage of old Sir Harry Rawson. Just *think* what risks he ran – the slightest hitch, the slightest delay, could have been fatal. But he accepted the risk, and it proved worth taking. But only just. Two days after the party got back on board, 99 per cent of us had West Coast fever.'

Gilbert Stephenson spent his last year as a Midshipman under sail, first in HMS *Active* and then in HMS *Raleigh*.

Most of the people on board were boys – there were only 15–18 Midshipmen. Stephenson was Midshipman of the Main Top in *Active*: this meant perching halfway up the mainmast in every kind of sea, supervising the setting and furling of sail. Like everything else on board, this was a highly *competitive* business, and when things got specially keen, the Midshipmen used to go out on the yards themselves and lend a hand:

'One thing that really astonished me at this time was the way drunken men used to go aloft: I remember seeing one Leading Seaman running up the rigging (we always ran up the rigging) – he was *very* drunk – and he ran out to the yardarm, rocking from side to side, but never a fall. How they did it, God knows – I think He must have watched over them! But they were wonderful men, those Petty Officers and Leading Hands – and most of them could neither read nor write. Were they uneducated? No. They were masters of their job and they did it. They were looked up to by the officers as well as the boys and men under them.'

Many of the boys on board were dreadfully seasick, and they used to be pretty roughly treated. They were given great troughs to be sick in, and after a very short time, were driven back to work. 'It sounds brutal, but perhaps it's the best way. I myself have always been lucky about seasickness; it must be dreadful for those who do suffer from it – one of my staff officers at Tobermory just could not get over it. But it never stopped him working. He used to go to sea almost every day in those little corvettes, and every day he was sick: I admired his endurance.

'Yes, those early days under sail were tough. I have never ever worked so hard, and I have never enjoyed life as much as I did then! It was wonderful experience, all that work with ropes: if you didn't do your duty, someone else could easily get killed, or break his bones. It taught us all *responsibility* – and that is something I have always enjoyed.'

# 3

# In Sole Charge

(Sub-Lieut. and Lieutenant, 1898–1903)

After this spell under sail, Gilbert Stephenson was duly promoted Sub-Lieutenant, and before long was off to sea again in HMS *Mermaid*, a destroyer of the First Flotilla (there was only one) based at Chatham.

Destroyers were fairly new to the Navy at that time – the first of them, HMS *Havock*, had been launched in 1893. This new race of fast, small warships had been developed to combat the threat of the torpedo – then carried and launched from swift torpedo-boats.

Speed, therefore, was of the essence in these 'Torpedo-Boat Destroyers', to give them their full name, and Stephenson's Captain in *Mermaid*, Commander J.M. de Robeck, who was later to rise to high office, exercised it to the full.

They used to have fun working up and down the Thames at speed, though they were really a menace to the sailing barges, especially when they had their hatches open. When the bargees saw the destroyer approaching, they used to make a dash to get the hatch-covers on; and a good many sharp words were exchanged as the vessels passed. De Robeck almost always got the best of these rallies. He passed within a few feet of one barge with an advertisement for Beecham's Pills on the sail, whose master let fly a tremendous torrent of foul language; but *Mermaid's* Captain had the parting shot. 'Go and lick your —— mainsail,' he yelled, 'and give your liver a chance!'

These were the days when 'spare the rod and spoil the child' was the theory behind naval discipline, and de Robeck certainly seems to have followed it. There was one particular signal boy he always picked out for savage treatment – and Sub-Lieutenant Stephenson often used to wonder why. One day the Captain literally kicked the lad off the Bridge – and then came the explanation: 'A bright fellow that – one day he will be a good signalman!' In his own terms, de Robeck had realised the boy was worth taking trouble about and was doing his best to encourage him.

While he was with the First Destroyer Flotilla, Stephenson was sometimes transferred to other ships; for a while he was First Lieutenant of a new destroyer in Chatham Dockyard – his first taste of full responsibility.

It was the depth of winter, and the unheated messdecks, were desperately cold. So the Sub-Lieutenant had what he *thought* was the bright idea of installing a pair of nightwatchman's braziers; stoked up well with coke, they solved the problem of warmth admirably. The problem of ventilation, however, remained, so Stephenson was most careful to issue orders in writing that the hatches were to be left open: 'I was in sole charge of the ship at that moment, so imagine my feelings when at about 6 a.m. the first morning the Coxswain rushed into my cabin shouting: "Half the crew are dead!"

'I ran forward to the messdeck and, for one terrible moment, I thought many of the men *were* dead. Suddenly I realised what had happened; they were gassed by fumes from my braziers! The well-meaning but silly Quartermaster had decided everyone would be even warmer with the hatches closed and had closed them – quite against my orders. How lucky for me that my orders on this subject were written!'

Sub-Lieutenant Stephenson managed to get a message immediately to the hospital, which fortunately was close at hand, and meanwhile frantically began looking up 'What to Do with Gassed Men' in the first aid book. But nature got to work pretty quickly and, with the help of plenty of fresh air, soon put everyone to rights. Not a very happy experience for a young Naval Officer in charge of a ship for the first time.

However, when trouble threatens, a ship's company can be

extremely loyal, as another story from these First Flotilla days makes clear.

Stephenson was on the bridge, again as First Lieutenant, when his ship was in collision with another destroyer. The Captain was in charge at the time, so there was no question of Stephenson being held responsible; all he was required to do at the Court of Enquiry was to answer detailed questions. However, the ship's company evidently thought he might be in difficulties, for on the morning of the enquiry the Coxswain came to the wardroom while he was eating his breakfast.

'Please, sir,' he said, 'the ship's company wants to know what you'd like them to say at this 'ere enquiry.'

'Coxswain,' replied No. 1, feeling both touched and amused by what he had been told, 'I should like them to tell the truth, though, quite candidly, I doubt if they know how!'

They were clearly prepared to go to any lengths to get him out of trouble.

Stephenson's next appointment was in sharp contrast to small-ship life. He went out to join the battleship HMS *Ramillies*, wearing the flag of Rear-Admiral Lord Charles Beresford, on the Mediterranean station. This was in the first year of the twentieth century.

At that time scarcely anyone even considered the possibility of war – after all there had been no major conflict involving Britain for nearly a hundred years. Everyone assumed that the 'Pax Britannica' would last for ever.

The Royal Navy followed time-worn, apparently unalterable routines. At sea, the Fleet kept up an endless round of so-called 'Equal-Speed Manoeuvres', which may have looked elegant, and tested the accuracy and judgement of the Captains, but had little relevance to fighting a battle.

At times when the Fleet was at anchor during the periodic cruises, the picket-boats were fitted out with masts and yards for signalling purposes and carried out their own Equal-Speed Manoeuvres. Each was commanded by a young Lieutenant (Stephenson was one of them) with the help of a Midshipman and Signalman. When the time came to return, the boats were formed up in line abreast and, on the 'Finish' signal being hauled down, each had to race back to its own ship: 'My goodness, what competition there was between us, for each picket boat represented its ship: and it was fine exercise for engine room staff, as well as those on deck. I well remember the splendid brass funnel we had made on board for my boat, and how we polished it!'

Competition was the mainspring of the Fleet in this time of peace. Ships' companies vied fiercely with each other in manoeuvres at sea, boatwork in harbour, smartness on parade, the appearance of their ships – everything, in fact, but preparation for war. Gunnery practice was reduced to a minimum for fear the enamel on the guns got spoilt.

All this was as true of the Mediterranean Fleet as it was of any other – until a very remarkable man arrived on the station as Commander-in-Chief in July 1899. This was Admiral Sir John (Jackie) Fisher. He was convinced that war would come in the fairly near future and, as things looked then, France seemed the most likely enemy.

In popular eyes this fearless man came eventually to acquire a reputation as perhaps our most remarkable sailor since Nelson. He was also a highly gifted if controversial administrator, and was to come into sharp conflict with Lord Charles Beresford in the first decade of the twentieth century. But in these Mediterranean days Fisher's qualities were just those which would attract an ambitious young Lieutenant – his 'prescience amounting almost to second sight', and his forceful way of expressing himself.

Under Fisher, Equal-Speed Manoeuvres vanished; the Fleet now proceeded in single line, in close order, which was presumably the way the C-in-C intended fighting the ships if war came. He created imaginary war situations to test the reactions of the men, he got the Fleet for the first time for a century to *think* in terms of war, he stimulated people's ideas.

One of his schemes was an essay competition among officers of the Fleet. There were two subjects – 'The Best Tactics for the Main Fleet' and 'The Best Stationing of Lookouts'. There was no compulsion to take part, but a large silver rose bowl was offered as first prize, plus an invitation to dinner for each competitor from the C-in-C, so the entry was pretty large: 'In this way Fisher got the help of the best brains in the Fleet at no cost to himself; compulsion alone would not have achieved his object.'

Sir John was a great destroyer of time-honoured myths of every kind. All engineers at that time seemed to have been brought up in the belief that to steam at full speed for more than three hours would damage their boilers beyond repair; so, when the C-in-C announced his intention of steaming at full speed all the way from Gibraltar to Golfe Juan, near Nice, there was a concerted protest. The Chief Engineers of all the ships got together, asked to see the Admiral and told him his orders would ruin their engines.

Fisher listened to them and then replied with characteristic bluntness: 'Gentlemen, if you can't do it, there are many engineers ashore at Devonport, Portsmouth and Chatham who will be only too happy to take your places. Good morning!'

The passage was duly made at full speed, and no damage whatever resulted.

Needless to say, Lieut. Stephenson's admiration for this great man was unbounded, and by a lucky chance he was soon in a position to see quite a lot of him. As one of his many measures in preparation for war, Fisher commissioned all the small vessels in the Reserve Fleet at Malta, and Stephenson was given his first command – of a 22 knot torpedo boat, TB 90. Captains of each and every ship – even torpedo boats – had the right of direct access to the C-in-C and the Commanding Officer of TB 90 took full advantage of it.

Quite frequently he used to send the Admiral a signal: 'Should like to see you when convenient' – and quite frequently it was convenient, so the Lieutenant came to spend much time in the company of a man who was, quite frankly, a hero to him: 'He saw what was coming when others were – or chose to be – blind. His single thought was *efficiency* and preparedness for war; and, later on, when in a position to do it, he recalled all our many small ships all over the world engaged in showing the flag (and often keeping the peace). By then, Germany had shown her intention to challenge our command of the seas and, in Fisher's opinion, we could not afford to dissipate our strength. We had to assemble a fleet sufficiently strong to meet and beat a rapidly growing enemy.'

Admiral Fisher must have recognised the qualities of his young admirer, for soon Lieutenant Stephenson, at the age of 23, was given a second command – this time of a much larger ship from the Reserve Fleet, the destroyer *Scourge*. However, their Lordships had other ideas. Stephenson was ordered to come home and join the next torpedo course at Greenwich. He naturally didn't want to leave the Mediterranean just then, but the move was to be significant for him in more ways than one.

From a professional point of view, it foreshadowed his future concern with underwater warfare and brought him into what was, at that time, the most rapidly advancing department of the Navy (the other three fields of specialisation for executive officers were gunnery, signals and navigation). And in the spring of 1902, while he was still on the course at Greenwich, something very important happened in his private life.

He was staying with an uncle and aunt of his at Frensham in Surrey and, one morning, an invitation arrived for Gilbert and his brother to ride over to Churt, not far away, to visit Lady Chesney, who had two young granddaughters with her: 'With the elder girl, Helen Chesney Williamson, I fell immediately in love and, curiously enough, so did she with me. But Helen was only seventeen and, when her father heard of what he considered this most unsuitable attachment to a penniless naval Lieutenant, he demanded that she should return instantly to Ireland.'

It seems there followed one of those rare but perfectly genuine cases of illness caused by frustrated love. Such was Helen's state of health that her doctor said he could take no further responsibility if the girl's feelings were disregarded. And so, to his great joy, Gilbert Stephenson was invited to spend his Christmas leave of 1902 with Helen and her parents: 'The next year we got married, and it was the wisest thing I ever did in my whole life. (It was just as well I acted quickly because I discovered afterwards that there were nineteen other people who wanted to marry her!)

'But I remember the senior Midshipman on my first ship saying to me: "Stephenson, never forget that you have joined *the* Service. It must always have first call on you. When that is fulfilled, *then* you may think about your family."

# 4

# 'Torps'

### (Lientenant and Lieutenant-Commander,
### 1902–1912)

Happily for the young couple, Lieutenant Stephenson was posted near home during most of his engagement and the early months of his marriage – as an instructor at the Torpedo School at HMS *Vernon* in Portsmouth. During that time he went to sea for only one short period of six weeks, on channel manoeuvres in the old battleship *Benbow*, and he remembered very little about those weeks except longing to be home again – and the hard business of 'coaling ship'.

This was a chore which came round every few weeks in the Navy of those days. All hands on board were employed on it, and gruelling work it was, filling the heavy sacks hour after hour and hoisting them up and over. Stephenson was in charge of the work in one hold, and the ratings employed there were almost all boys: 'After twelve hours of almost unceasing work I could see they were utterly exhausted, so I asked the Commander to stop coaling for the night. He refused, saying he had the Captain's orders to go on. "Then I ask to see the Captain," I said – which was a bit uppish of me, seeing that I was only the Torpedo Lieutenant.

"Certainly not," replied the Commander, "he's in bed." This did not put me off; I insisted on seeing the Captain, so the Commander took me in and, to my great relief, the Captain decided to call off the coaling operation for that night.'

Almost every part of life in the Navy of those days was ruled by the spirit of competition, and coaling ship was no exception. As a Divisional Officer in HMS *Monmouth,* a new three-funnelled first-class cruiser to which he was appointed at the end of 1903, Lieutenant Stephenson made it his business to see that his foretopmen beat the forecastle, maintop and quarterdeck divisions in coaling as in other matters.

The upper deck party was responsible for rigging derricks for hoisting ten or twelve coal bags at a time up from the hold of the collier and over to the deck of their own ship, while the engine room department were responsible for the exhausting and unpleasant work of trimming the coal as it landed at the bottom of the shute into the bunker.

Not only was there competition between divisions in coaling; there was sharp rivalry, too, between the different ships of the Fleet. As the job went on, each ship would signal the number of tons loaded in an hour – and the speed could be almost unbelievable. Once, at Gibraltar, they took 1,100 tons of coal on board HMS *Black Prince* from lighters in less than three hours, at a rate of no less than 400 tons an hour! 'You can imagine how hard the men worked to achieve that kind of speed; they had to be tough to endure it. Once when we were coaling I noticed a seaman with bare feet working in the hold and told him to go and put his boots on. "No fear," said the young man, "I don't get my boots spoiled on this ruddy coal!" In fact, in those days seamen hardly ever wore boots except on Sundays and for going ashore, so their feet were pretty hard.

'The one thing to be said for coaling was this: there was nothing quite like it for sorting out the good men – the men who had good hearts – from the weak ones.'

One might think the men would hate this hard and dirty work, but apparently the element of competition made it exciting. It all began with the pipe 'Hands to *clean* into coaling rig' (an odd way of telling people to get into their oldest and scruffiest clothes). Then, on the second pipe, 'Hands fall in for coaling', they all came rushing out. When all were reported present to the Commander and he had said a few words to them, he snapped out the order: 'Coal ship!' All turned and simply rushed to their stations, as the time was taken from the words 'Coal ship'.

It was undoubtedly an honour for Lieutenant Stephenson to be appointed Torpedo Lieutenant in HMS *Monmouth* – she was a new ship, fitted with all the latest electrical gear, which was the responsibility of the Torpedo Department. Some of this gear was not very efficient; for instance, *Monmouth* was one of the first ships to have electrical boat hoists, and on her trials the dock-yard people were quite unable to get them to work. However, 'Torps' thought he could do something with them and agreed to take them over: 'Luckily I did manage to hit on a solution and, although the motors produced the most alarming flames whenever they were used and an electrical artificer had to stand by every time we switched on, there was no actual *failure* during the two years I was in charge.'

They were also concerned with early wireless experiments in *Monmouth*. These were very early days for radio, and naturally all communications were in Morse code: 'You can imagine how staggered I was when one day the Radio Petty Officer on watch came to me in a great state and said: "I heard a man *talking*, sir!" He thought he must be drunk.'

Lieutenant Stephenson's natural instincts found an outlet at this time through boxing and, as *Monmouth* was the first sizeable ship he had served in for any length of time, he had the perfect opportunity to help train a good boxing team. (His nickname, by the way, among his naval contemporaries was always 'Puggy', though perhaps this had less to do with his interest in pugilism than with his personal appearance!)

The Lieutenant found this leisure-time activity was a marvellous way to get on close terms with the men. Whenever he got the chance, right through his career, he used to run ships' teams; but he would never on any account handle the money– someone else had to be Treasurer: 'I have always been appallingly bad with accounts, so I would probably have made a mess of it and, as I said before, it's not enough to *be* honest, you've got to be *seen* to be honest!'

After *Monmouth* Lieutenant Stephenson had another two years at sea from 1906–1908 as Torpedo Officer of HMS *Black Prince*, a large armoured cruiser. She had six 9.2-inch guns and twelve 6-inch guns and was a very advanced ship for her day with a great deal of electrical gear including, for the first time, electrical training and elevation of the guns and electrical ammunition hoists.

With so many experimental features, especially in the gunnery department, it is not surprising that things went wrong from time to time. There was no drill laid down as yet for correcting the faults that arose, so a good deal of perception and initiative was required of the torpedo department.

The leads which carried electrical power into the moving parts of the gun turrets were subject to great strain, and a set of them parted on one occasion. There was only one way to find out what had gone wrong – and that was for 'Torps' to insert himself down the very narrow trunking which carried these leads. Unfortunately he got stuck and no amount of pushing and pulling would free him: 'The Petty Officer was in a *terrible* state. But eventually I was extricated with the help of a large quantity of soft soap!'

Another nasty moment was when Gilbert Stephenson was very nearly seasick for the first and only time in his life.

*Black Prince* was not very well ventilated at the best of times, and during her steam trials the ship was completely shut down. The weather was bad and so was the smell; sitting in the Wardroom, the Torpedo Lieutenant began to feel very ill.

Suddenly a messenger rushed in – number 5 fan in number 2 boiler room had stopped working. Away went the seasickness; 'Torps' immediately put on his overalls and ran down. Conditions in the boiler room were bad and the fans were placed in the hottest parts of all. The men on watch down there had to wear the thickest possible clothes to protect them from the heat, and were never left to work alone in case they should faint: 'But as soon as I got the call to go down there, unpleasant though it was, my seasickness was gone. The need for action had banished it, as I believe it almost always will.'

It was in *Black Prince* that the Lieutenant first learned to treat dockyard promises with great reserve.

The Admiralty had decided to make the ship more efficient by fitting electrical fire-control instruments to relay range and deflection instructions to the turrets from a central gunnery-control position. The dockyard got so far as to install the wiring by the time the ship was due to go to sea again, but they had not begun to put the new instruments in.

'That's all right,' said the dockyard electrical officer to Lieutenant Stephenson, 'You can install the instruments yourself when you get to sea. The wiring is all correct.'

'What!' replied 'Torps', 'Without the chance of *testing* them first? I wouldn't *dream* of it!'

The Captain agreed to stay until the dockyard had fitted the instruments and an official trial had taken place. 'And sure enough, when it did, *nothing, absolutely nothing*, worked. The whole equipment was condemned.'

From what has been said so far, it might be imagined that a Torpedo Officer never had anything whatever to do with torpedoes. But of course, from the point of view of the ship's fighting capacity, this was the central part of the job.

The torpedo rooms in *Black Prince* were below the waterline, and this fact created special risks which were met by an elaborate system of safety precautions. Lieutenant Stephenson knew that precautions were needed, but had also come to the conclusion that, if they were all observed to the letter under war conditions, his rate of firing torpedoes would be quite inadequate. So he decided to revise the torpedo-launching drill.

In the torpedo room, the heavy torpedoes were moved from racks on the bulkheads to the torpedo tubes themselves by chain pulleys running on overhead tracks, and then loaded into the tube through an inner door which opened into the room. Needless to say, this door and a side door had to be firmly closed before the 'sea door' at the outboard end of the tube was opened, thus flooding the tube, with the torpedo inside ready for firing.

The established drill laid down many precautions to ensure that the 'sea door' would not be opened too soon, with obviously catastrophic results. Although there was an element of risk, Stephenson cut these precautions down to a minimum and rehearsed the men in his new and speedier drill. It worked well – provided there was a high degree of concentration and efficiency from all concerned.

'Torps' was so pleased with what had been achieved that he invited Admiral Prince Louis of Battenberg, father of Lord Mountbatten, to come and watch his team at work. Alas, the worst had to happen. Someone who was not on his toes managed to open the sea door *before* the inner door was closed!

'You can imagine the result. The sea simply flooded in through the vast 21-inch hole below the waterline. In no time, the torpedo flat was two feet deep in water and all of us – Admiral included – were soaked to the waist. Thank God, we managed to get the door closed and the Admiral up the escape.

'Now I was obviously very much at fault and Prince Louis could have made things very difficult for me. But, fortunately, he sympathised with what one had been trying to do – raise the rate of firing and thus increase the efficiency of the ship – and he never reported this bad incident to my Captain. I admired him greatly for that – and not only for that.'

Stephenson had very little time at home during his four years in *Monmouth* and *Black Prince*, though he wrote to his wife every single day; so, on paying off, he welcomed the signal which appointed him to the Royal Naval War College at Portsmouth, where he was a member of the war course staff from August 1908 until August 1910.

During this period he became a Lieutenant-Commander and helped organise the elaborate war games which provided an opportunity to try out various tactics and assess the rival merits of gun or torpedo fire as a means of attack. The only trouble here was that the results of imaginary *torpedo* fire were so much more difficult to establish than the effect of gunfire in any given situation – in fact hits or misses were decided on the war course with a bag of brown and black marbles! If you had 'fired a torpedo', you put your hand in the bag and if you pulled out a brown marble you had missed: 'There was one Admiral, I remember, who used to cheat by warming a black marble, so that he could produce a hit every time he "fired"!'

Torpedoes were not taken very seriously as weapons of war at that time. Admiral Lewis Baily, who was Admiral President of the War College, told the course that torpedoes were simply one of the risks of war: by a strange irony, one of his ships was torpedoed in the English Channel very soon after the First World War began.

On the whole, it seems, this war course was very well run and covered a wide range of subjects, including economics. But it was not really enough: 'Time was running out and the country desperately needed a highly trained Staff Corps; we had nothing remotely like it.'

After War College, Lieutenant-Commander Stephenson had two and a half years in the battleship *Duncan* (wearing the flag of Rear Admiral Thomas Jerram, who was to command the Second Battle Squadron at Jutland) – the first two years as First and Torpedo Lieutenant.

Being in the zone for promotion to Commander in those days meant living a very busy social life – you simply had to be seen at the right parties – and this made great demands.

Gilbert Stephenson took the opportunity of living ashore with his wife when the ship was based at Malta, and they made a determined attempt to go to as few parties as they could. Even so, they went to fifty in the course of one season. It was quite usual to arrive home at 2 a.m. and then be up again at five to get on board by 6.15!

Sometimes, of course, there would be a dance on board *Duncan* – and her officers tried to think up something new every time this happened. And, as *Duncan* won most competitions in the Fleet, her dances had to be the most successful too.

They did rather well with their lighting. The quarterdeck used to be covered with a red and white awning for dances and the First Lieutenant managed to concoct a paint for the light bulbs which exactly matched the red in the awning. This had completely defeated the other ships in the Fleet – the ordinary Admiralty red was a kind of magenta and didn't match the awning red at all: 'You can imagine the attempts that were made to get the formula; I am glad to say they were unsuccessful!'

In those days, officers were expected to spend their own money on extra touches of elegance for their ships: 'I don't remember how much I forked out on coconut matting and enamel, but there was a great satisfaction in knowing the appearance of your own ship was the best.'

Apparently in those days there was quite a flourishing black market for the sale of ship's paint ashore in Malta, and one day *Duncan's* painter came to the First Lieutenant and told him he'd had an offer for substantial quantities of paint from the ship. 'No. 1' told him to agree to the proposal – and to inform the

Commander at what time and on what night the paint would be picked up from the ship. Then, of course, it was arranged to have a floodlight ready to be focussed on the culprits as they lay alongside, and a picket boat lay at the boom with steam up to pursue them ashore.

They were caught red-handed, but the ship's painter had to be sent back to England, and his name changed, too!

The First Lieutenant greatly admired his senior officers in *Duncan* – Rear Admiral Thomas Jerram and Captain Lawrence Field – indeed they were generally popular. One day at Waterloo Station a stoker from the ship came up to the Admiral in great excitement and said: 'Sir, yesterday the Watch got in 152 tons of coal an hour!' Admiral Jerram was delighted. The stoker had not hesitated to tell him of their achievement, and clearly knew he would be interested.

Captain Field, besides being a first-class officer, was an adept conjurer, especially with cards, and this was a great asset when the ship was entertaining foreign officers. He also became godfather to the Stephensons' son Gilbert.

Promotion came again in *Duncan* and for his last few months on board, Gilbert Stephenson was Commander of the ship – in fact for three of those months he was actually in command, for after *Duncan* had become a private ship, the Captain was called away to advise the Admiral in a protracted collision case. This was a remarkable opportunity – and a remarkable tribute to the abilities of so very junior a Commander – in command of a battleship at the age of 34.

# 5

# Able-Bodied Young Commander

## (1913–1916)

In March 1913, Commander Stephenson was appointed to the Intelligence Division at the Admiralty with special responsibility for the waters of North and South America. He and his colleagues had two periods in the War Room during manoeuvres when the Reserve Fleet was called up, and Stephenson was there when war was declared in August 1914 – the first major war for a hundred years. The downright qualities of 'The Terror of Tobermory' began to assert themselves: 'The system in the War Room at the start was very inadequate. None of the Commanders on duty there was able to take action without the approval of a Post-Captain. This was quite ridiculous, and I said so! My suggestion for more delegation of authority was not approved.

'But when I worked as a Duty Commander – 24 hours on, 48 hours off – there was no question of referring everything to senior officers. The messages that came in had to be acted upon – and of course that suited me much better.'

The Commander was not unhappy at spending these first months of the war in Whitehall. The War Office was practically evacuated, on the assumption that the war would last only six weeks or so; but it seemed silly for an experienced man to press for a sea appointment until things had settled down a bit. Then, after five months of war, he was told he could go if he found a suitable relief.

So he started to look around – 'After all, what's the point of being at the Admiralty if you can't find yourself a good billet when you leave?' He decided on command of a special force of P Boats that was being set up. In this state of mind, he was summoned before the Fourth Sea Lord and given orders to go to Alexandria and reorganise transport arrangements there!

Commander Stephenson didn't like that idea at all. 'I thought if my children were to ask me: "What did you do in the war, Daddy?" I would have to reply: "Well, I started with a comfortable armchair at the Admiralty and went on to another armchair at Alexandria." No, I couldn't bear the thought. So I went to the Director of Transports.

"Look here," I said, "do you *really* want able-bodied young Commanders in your department?"

"No," he said.

'That was good enough for me. I went to the officer who made Commanders' appointments. "Have you got *too many* able young Commanders for sea duties," I said, "that I should be sent to Alexandria?"

'That finished it. I never heard another word about Alex. and the transport appointment. (The lucky thing about this was that no one ever found out how the appointment came to be altered!)'

Soon afterwards, in the spring of 1915, the Naval Secretary to the First Sea Lord sent for Stephenson and told him that Vice-Admiral Sir John de Robeck (his former Captain in *Mermaid* and now in command of the Dardanelles operation) wanted five young Captains or Commanders to help on the beaches at Gallipoli, and he asked him if he would go.

There was only one answer to give him.

The other four officers were sent to Gallipoli overland, but Stephenson was sent out by sea in charge of 500 Naval New Entries in the *Franconia* – which was also carrying 1,500 of Kitchener's Army.

This voyage presented problems. The Commander was very concerned about the effect on his untrained men of three weeks' idleness, cooped up in a ship. So, as a preliminary step, he got hold of all the gymnastic gear he could find which was suitable for use on board.

The *Franconia* party included twelve newly commissioned RNR Sub-Lieutenants, and, fortunately, a Chief Petty Officer and twelve Petty Officers, 'all of them seasoned men and the salt of the earth'.

Stephenson's first concern was to get his men separate quarters – he had to conceal their total ignorance of the Navy and its traditions from the soldiers! Then he had to start keeping his men busy and out of trouble.

The first time they sat down to a meal, there was complete chaos. Afterwards, the Chief came to the Commander's cabin. 'Sir,' he said, 'they're like wild animals. Only the men nearest the food got anything at all – three-quarters of them got nothing.'

So, at the next mealtime, the Commander went to the messdeck himself. 'You will all stand up and remain standing,' he said. 'No food will be issued till I give the order. When I do, the food will be passed straight to the far end of the table. You will still remain standing and *no one* will start eating till I tell you to sit down.'

The Commander waited until the food had been properly distributed, said 'sit down', and left the messdeck.

The next thing that happened was a deputation from the men saying they didn't want porridge for breakfast – it was only fit for the hounds: 'Well, I know some people don't like porridge, so I asked the Captain for something else. This incident had a curious bearing on the question of moral courage, though I didn't realise it till later when selecting men for a working party on the beach at Gallipoli. The fittest men were wanted, so I made a quick medical check by looking at the state of their feet and mouths, and do you know that among those men who had thrown their porridge overboard, there were twelve with no teeth, top or bottom. Porridge was about the best thing they could eat. But when all their mates started saying they didn't like porridge, they hadn't the guts to say they did!'

There was also a deputation in the *Franconia* asking for a spirit ration.

Now the Commander didn't really want to issue a spirit ration – the men had quite enough spirit in them as it was. In any case there was no rum to give them. But the men said there was a notice up in the ship saying they were *entitled* to a spirit ration. ('This was a very great mistake in my view, but there it was, in black and white – the men were quite justified.') So Stephenson went to the Captain and asked for a daily issue of spirit for the sailors.

'But I *have* no rum,' said the Captain.

'That doesn't matter,' replied the Commander, 'the notice refers to a spirit ration. Haven't you got any whisky or brandy on board?'

'Now look here,' said the Captain, 'do you seriously expect me to start issuing brandy at the price it is nowadays?'

'Well, if you had not put up a notice stating that Naval ratings are allowed a spirit ration, this trouble would not have occurred. But, as it is, it is my obvious duty to see that my men get what is due to them.'

The Commander insisted that something would have to be done, although the *Franconia*'s Captain was very upset at the thought of what his Company might say to him about it.

'Now look here,' said Stephenson, 'if you will make a call at Gibraltar I will send a signal to the Admiral there, asking him to send out some rum to us with the implements to serve it.'

So that is what was done: 'I regretted it, mind you. But you see, although it was my job to stop my men doing wrong, I also had to make sure they got their rights!'

There was something the Commander considered much more important than a spirit ration, and that was exercise.

He arranged for boat drill four times a day; and, with the idea of giving his sailors a bit of professional pride, he asked that some of them should be allocated to each boat. The GOC agreed that the Naval 'experts' should have priority in these drills over the Army: 'The cry was "Make way for the Sailors!" – and that helped morale. Thank God their "expertise" was never put to the test!'

The General undertook to have one deck left completely clear after the midday meal, and the vigorous Commander used to fall his men in five deep and just keep them running round the deck till exhausted. This was not very popular – hardly to his surprise – and it led to another deputation. Some of the lower deck lawyers pointed out that it was not customary for men over twenty-four years of age to do physical training. Commander Stephenson was quite equal to this. He pointed out that the Turks were unlikely to know about this custom, and would not make allowances for it when in full pursuit with sharp twisted bayonets. However, he did make a show of compromise in the matter. 'I am prepared to meet you half-way,' he said. 'All men over forty are excused PT.' Of course he knew perfectly well that none of them was nearly forty.

There were, of course, some disciplinary problems with all those men shut up in a ship for so long, and one day the Chief Petty Officer reported a bad fight on one of the messdecks.

'Right,' said 'Puggy' Stephenson, resorting to a well-tried remedy, 'we'll have a boxing meeting tonight!' He got the two men facing each other in the ring and announced a fight to the finish. There were no more messdeck scraps after that.

By the time the voyage was over, in the Commander's view at least, the members of the Naval contingent had reached 'a happy understanding'. When they reached Gallipoli there was no shortage of volunteers among the men for dangerous jobs ashore – though such was their loyalty to 'Puggy' that many refused to leave the ship without his direct orders. However the Admiral in charge there, Admiral Wemyss, took this well; he merely remarked: 'You've evidently got a good grip on your men!'

There was one particularly ferocious officer at Gallipoli – a Captain Carver, who had the reputation of provoking more mutinies than anyone else in the Royal Navy. He was supervising salvage work in the tug *Alice*, and his crew were terrified of him – much more than they were of the Turks, though they were working under Turkish fire most of the time.

One day Commander Stephenson had to draft fifty men to work under Carver, and the next morning Admiral Wemyss told him that Carver had had yet another mutiny and was under arrest.

'Right,' said the Commander, 'would you like me to settle this mutiny business myself, and no more said?' 'For God's sake do!' was the reply. So Stephenson told Carver the Admiral would let him off this time and that he would allow him to have another working party, but if they were not treated in a more civilised way the Admiral would send Carver home to be discharged – Not Required in War.

Next day a fresh party of fifty men was sent to Carver under one of the RNR subs. They got back on board at 2 a.m. After the Commander had sent them down for some food, he called the Sub. to his cabin.

'Well,' he asked, 'how did you get on? I asked him to be gentle.'

'Did you, sir?' replied the Sub. 'In that case, God help me if I ever meet Captain Carver when he's not gentle!'

The Turkish capital was dominated by the Germans, and Turkey was in a position of the greatest strategic importance, effectively blocking supplies through the Black Sea to our Russian allies and preventing the export of Russian grain. So everything possible had to be done to undermine Turkish power in the Eastern Mediterranean.

Stephenson was appointed second in command of a very old battleship, *Canopus*, which was guard ship at the Aegean island of Mytilene, and at first he was disappointed to be sent to such an antique. But in fact the job turned out to be most interesting.

*Canopus* was Headquarters ship for a very big area; she had a fleet of motor gun-boats attached to her-all manned by the RNVR – and a number of French airmen, to keep watch on the Smyrna Gulf. One of the jobs was to try to distract the Turks away from the Russian front by making them suspect an attack on Smyrna.

These early planes were very shaky articles and it needed a good deal of courage to fly them. The French boys had plenty of it. However impossible their orders seemed, they would just shrug their shoulders and say: 'Pourquoi non?' They were characteristic in other ways too. The Commander once removed a lady he strongly suspected of spying from a cargo ship, and a few nights later, when he was dining at the French camp, she was introduced to him.

'But haven't we met somewhere before?' he asked. She was now happily employed in a different role with the French airmen.

Commander Stephenson's own attitude to flying was not as courageous as theirs. He went up once in the only British plane they had, on the only mission it ever flew: 'I remember it was dreadfully difficult to start the thing and our rise into the air was so unsteady that I shouted: "For God's sake look where you're going!" We managed to get to the Smyrna Gulf and dropped two bombs, though I don't think we hit anything. After that the plane never flew again.'

Another task was to land and recover secret agents in the Smyrna area, a job greatly relished by the Royal Naval Volunteer Reservists serving with *Canopus*.

This, in fact, was Gilbert Stephenson's first contact with the RNVR and he soon began to admire them greatly. Most of them were stockbrokers, barristers, doctors, architects and so on, and had scarcely been in the Navy five minutes; but in their yachts and motor gun-boats they would go right ashore at night under the Turkish guns with the greatest courage and gallantry. They seemed to love this risky work: 'Mind you, they were quite undisciplined-there was no telling what they would do next; you might give them a certain job to do and they would be quite likely to shoot off at right angles if they saw something of marked interest elsewhere!'

One of the more interesting episodes at this time was a raid on the Greek island of Chios, where a group of five German and Turkish troublemakers were active. Greece was neutral, so the Navy hadn't really the slightest right to interfere – and Chios had a garrison of 500 Greek soldiers! The Commander was in charge of this, to say the least, delicate operation.

He had asked the British Vice-Consul there to provide accurate and detailed plans of the island, marking the best routes from the quayside to the houses where the malefactors lived. Five separate raiding parties were formed, each under the command of a young officer and including a carpenter with housebreaking tools, and two Marines with sacks to carry off any papers that were found. To back them up, the Commander had a party of 100 armed seamen who would stand by at the quay in case of trouble. But the great aim was to accomplish the mission with the minimum of turmoil.

The party left Mytilene after dark in a corvette, which brought them off the island of Chios at 2.30 a.m. A drifter met them outside and took the landing party into the harbour.

All went exactly as planned. Once they were ashore, the Commander despatched the five parties, each charged with the arrest of one specified individual, and sent a polite note to the Commanding Officer of the Greek garrison telling him not to worry, the British had only come to remove several 'disturbing characters' from the island! ('What *would* have happened if the Greek soldiers had done what they should have done and set about removing *us?* Fortunately they did nothing!')

Three of the wanted men were found and arrested with the greatest of ease. The fourth was spending the night with his girl-friend; he was traced quite quickly, but not unduly hustled.

The only one to give serious trouble was the German Vice-Consul who threatened to fire on the British party from the upper window of the Consulate. The Commander went there at once, with the British Consul, who was acting as interpreter and guide.

'If you do not open the door before I count ten,' the Commander ordered the Consul to say in German, 'I will break it down with dynamite. This may well blow the whole house down, and you will be responsible for the death of all the women and children in it!'

In fact the Commander didn't have a scrap of dynamite. But 'fortunately wisdom entered into the man's head and the door was opened'.

The old German consular shield was hung in the hall of Sir Gilbert Stephenson's house – as a souvenir of that night's adventure on Chios. It was a complete success – and not a soul was hurt on either side.

# 6

# Uncrowned King

(Commander, 1916)

In March 1916, not long after the success of the Chios raid,
Admiral de Robeck visited *Canopus* and invited Gilbert
Stephenson to take over a key job as Commander, Crete Patrols.
This was every schoolboy's idea of an adventurous war
assignment.

At that time Crete, like mainland Greece to which it had been
ceded in 1913, was neutral – but, as in Greece itself, opinion
was sharply divided between those loyal to King Constantine,
who favoured the German side, and supporters of the liberal
statesman Venizelos, who wanted to bring Greece in with the
Allies.

Stephenson's job was to prevent any contact between Cretans
and the Germans, and to reinforce the rising tide of support on
the island for Venizelos and the allied cause. For this formidable
task, his vessel was an eight-knot trawler, like the majority of his
flotilla. The crews-and most of the captains-were quite illiterate,
but they were superb seamen – 'their powers of observation were
astounding – they simply *knew the way!*'

One day the British Consul-General said he had most reliable
news that a Turkish fishing boat had agreed to rendezvous with
a German submarine in a certain lonely bay by night. This
sounded a grand opportunity for the Crete Patrol.

The Commander asked a young RNVR Sub. in command of
a trawler to take his vessel to the bay after dark and find the

fishing boat and its crew. He was to take them on board his trawler, and, with a companion, to take their places, dressed in the Turks' clothes; to take a lance-bomb and lie in wait for the submarine – then to endeavour to get on board with the bomb and drop it down the conning tower.

The Commander made rendezvous with the trawler next morning. The Sub. reported all had gone well in finding the boat and dressing up in the fishermen's clothes, but a steamer had nearly run them down and, to avoid it, they had had to cut the fishermen's nets. No submarine had appeared. Meanwhile they had two Turks still on board the trawler.

The Commander had the Turks sent over to his ship, handcuffed and bound. Then he questioned them fiercely about what they had been doing in the bay. They denied all knowledge of a German submarine. Determined to get something out of them, Stephenson had fire-bars attached to their feet and told them that, unless they told the *truth,* they would be thrown over the side. They continued to deny any villainy and eventually he decided they were quite honest! So he let them go, with £36 in their pocket for the loss of their nets.

At a nearby village about two weeks later, the Vice-Consul told the Commander that two Turkish fishermen had reported that they had been most cruelly treated on board one of his ships. The Commander protested: 'Of course you don't believe such lies!'

As the Vice-Consul was about to answer, he exclaimed: 'There is one of them!'

'Bring the liar here,' said Stephenson, 'and ask him how many funnels the ship had.'

'Two,' replied the Turk.

The Commander exploded indignantly: 'Not *one* of my ships has *two* funnels. Send him away before I lose my temper.'

There were a number of paid agents in the villages of Crete to keep the British informed about German contacts with the island and, in particular, to warn of the presence of German submarines; but the Commander strongly suspected both their integrity and their intelligence. So he asked if he could borrow a British submarine, which could perfectly well masquerade as a German one for his purpose – and which would enable him to test the reliability of his Cretan agents.

E.21 was duly assigned for this work, and they went round the island paying calls. The British Consul-General (whose German was good) agreed to dress and act the part of a German Lieutenant – with Commander Turle and Commander Stephenson as German sailors. The locals received them in a perfectly friendly manner and happily parted with quite a lot of information about British movements as far as they knew them. The 'German' Lieutenant thanked them and promised to return for more information in two days' time.

The three of them did return, landing in the submarine's rubber dinghy. They were very nearly caught by an armed party of their friends, who of course imagined they were Germans!

A few days later they returned in their true characters, in Stephenson's trawler, and went ashore to question their contacts about this 'German visit'. The reports they received 'bore *absolutely* no resemblance to what had happened!'

At this time there was a strong revolutionary movement in Crete, and word had got round that Venizelos was only awaiting news of a successful revolt before coming to the island himself to take charge.

This was the background to a message one day from a corvette lying off Candia that the hoped-for revolution had indeed broken out, that about 3,000 insurgents had taken over the town. Stephenson had orders not to assist them – but he knew that without help they would fail.

Crete was part of a neutral state – the British had no right to interfere in their affairs; and Cretan cooperation was of the greatest importance, so whatever he did, the Commander knew he must not antagonise the local people.

He set course for Candia at full speed – eight knots. On arrival a message was sent to the British Vice-Consul to summon to the Consulate at 09:00 hours all the allied Consuls, the Prefect, Chief of Police – and the leader of the revolutionaries (though instructions were given to place him in a separate room!).

You will notice that the 'summons', as you might call it, was sent by a comparatively junior officer of a foreign country; but that country happened to be Great Britain, and there was no greater – moreover the Royal Navy was her most respected instrument: 'Believe me, I had not the slightest doubt but that my summons would be complied with – and it was!'

When all were assembled, the Commander first of all called on the Prefect to give a report on the situation in his town. He stated that about 3,000 men had attacked and broken into the town and had overcome its defences; many had been killed and, with the exception of a few strong houses, the town was in the hands of the revolutionaries.

'And what are you doing about keeping order?'

'Nothing,' he replied.

He was obviously, thought Stephenson, *quite* incompetent.

The Commander then visited Mr George Maris, the revolutionary leader. He could not speak English and Stephenson could not speak Greek, so they conversed in French.

'I understand you are in command of large areas of this town. Where are your men now?'

'I don't know.'

'What arrangements have you made for feeding and lodging your men?'

'None.'

'What steps have you taken to prevent looting and the settlement of private feuds and so forth?'

'None.'

The Commander then pointed out that Maris was personally responsible for the lives of all in the town, as he had taken it – and asked him what he was going to do. He seemed very uncertain. Maris said that he had been ordered to start the revolution and that was all; he expected Venizelos to arrive and instruct him further.

'But action is needed at once,' said the Commander. 'You must get your men fed and patrols set up. Now suppose I get the government parties to undertake not to fire from inside the houses they occupy – will you get your men to promise not to attack any of these houses?'

Maris agreed; Commander Stephenson retired to the other room and told those present of the proposal. They also agreed. The British Vice-Consul and Stephenson then proceeded to the main government house. There was great difficulty in persuading them to open the door, until they recognised the

British uniform. When the Britons did get in, the occupants were delighted to agree to the cease-fire proposal, for the house was crowded with men, women and children, and could easily have been taken.

As the Commander got up to leave, a great uproar was heard outside. The insurgents had arrived. 'My God,' he thought, 'I'm too late!' He had visions of the most awful carnage.

There was only one thing to do.

'Open the door *wide!*' he ordered.

Surely the sight of all those women and children would make the mob think twice! But no – all the rifles went up to the ready. Luckily there was no nervous trigger finger, and as soon as the crowd caught sight of the British uniform all the rifles went down. Whereupon 'Puggy' cursed them loud and long in every language he could think of, which was mostly English, and told them to get the hell out of it – which, to his very great surprise and relief, they did.

All went well for a day or two. Then Maris, the revolutionary leader, came to the Commander and said his men were getting very restive because Venizelos had not kept his promise to appear as soon as the revolt took place. Maris felt many of his men were losing confidence in the movement and were ready to disperse; the only hope, he thought, of keeping the rebels together was if Commander Stephenson were able to guarantee them a safe getaway by sea if things went wrong.

Now he had only a little trawler, and not the faintest idea how he would carry out such a guarantee; but he knew the success of this revolt was of great importance to Britain, so he gave his word. And Maris went away totally happy; he had unqualified confidence in the word of an Englishman.

What could Stephenson have done if there had really been trouble? His crew were totally unarmed – just fishermen; he couldn't have landed a party if he'd wanted to. He had no real power of any kind, but he was a British naval officer, and that was enough. It had to be.

After another day or two, things were beginning to get really difficult in the town and, although Maris had done his best to follow Stephenson's advice and keep his men occupied, they were getting thoroughly awkward and out of hand.

'You are my last chance,' he said. 'You must take a walk round the town with me!'

Well, of course, this was the last thing Stephenson wanted to do; he had, in any case, been ordered not to interfere. But he decided it was best to do as Maris wished.

The effect was remarkable. Everywhere they went, people stood up out of respect for the British uniform.

But that wasn't enough for Maris. Two days later he begged the Commander to go and talk to the men; however hard he tried to keep out, he was getting more and more involved! He promised to go – if his visit was kept dark. Luckily, Professor Dawkins, the Greek language expert, was on hand, so he could interpret. They were welcomed by an enormous guard of about 200 men armed with every conceivable type of weapon – confirming the reputation for ferocity of the Cretan peasant.

With Dawkins to translate, the Commander stood up and told them squarely that the eyes of the world were on Crete; were they going to murder each other? – or were they going to revive the traditions of ancient Greece, when their country led the whole world? It seemed to go down well and Maris was full of thanks. 'So he should be,' thought Stephenson: he had visions of himself strung up somewhere, and his wife and children starving. And he was going against orders all the time!

But, however scared he was inside, he saw that what these people wanted to believe about the British was that we were devilish fierce and quite unbeatable: 'We had been sent some propaganda leaflets which made out that we were gentle and Christian and all that, and the Germans were wicked and violent; it was no use at all! I had the leaflets burnt and Professors Dawkins and Halliday concocted some marvellous stories of successful Britishers and simple frightened Germans, and these were spread about because I knew instinctively that the Cretans had to believe us to be wolves and tigers – they didn't want to be on the good side, they wanted to be on the *winning* side!'

Diplomacy was undoubtedly essential in this situation, but so, in the Commander's view, was downright determination.

Once he had to call on the very pro-German Prefect at Suda Bay, though the British Consul there had warned him he would not return his call, as etiquette demanded.

'Never mind,' Stephenson replied, 'I'll see what can be done!'

So he made his call on the Prefect and had a general formal chat. Then, instead of *inviting* him to return the call and risking a refusal, the Commander said, as he got up to leave: 'Now, *when* you return my call, I shall send my boat with a uniformed crew to bring you off from the jetty. *At what time* shall we be there?' He called all right!

The whole of Commander Stephenson's time in Crete was fairly hair-raising but his most terrifying experience came when things seemed to have settled down quite well and he was dining with the French Consul one night in Candia. Suddenly in rushed Maris with a tremendous gale of wind behind him.

'You must come at once,' he shouted. 'There's a huge crowd outside the house where the Government party are, and they want to break the door down. I can't do anything with them – there'll be a bloodbath!'

'But I gave you a company of soldiers to help keep order – what are they doing?' asked Stephenson.

'Oh, they're worse than the rest!'

What was the Commander to do? His orders forbade him to interfere, but once again his conscience would not let him sit still.

Maris gave him an escort and off they went, in the pitch dark. Luckily the Commander had a good torch with him, and that showed up his British uniform: 'When I arrived, I cursed them all at the top of my voice and I told them to send for the Chief of Police. (I was very angry – and very frightened.) As soon as he arrived, I told him to open the doors and let everyone out and, if anyone was touched, I would personally cut the miscreant's throat the next morning. Somehow it worked.'

Venizelos, in fact, never did come to Crete; instead, in September 190, he set up a pro-allied 'militant state' in Salonika. But the revolutionary movement took hold all the same in the island, and in 1917 Venizelos, restored to power in Athens, led Crete, like the rest of Greece, into the war on our side. In due course Commander Stephenson's contributions to all this was recognised. A special delegation was sent to him to express Venizelos' thanks and to present a signed photograph of the Greek leader.

During those months in 1916 it was Stephenson who had made himself the arbiter of Crete's fate – as this little anecdote bears out.

Thanks to the efforts of Maris and his friends, and the general success of the Venizelist cause, the Commander eventually had the job of landing a new Governor at Candia.

Just as they were coming in to the harbour, the Governor said: 'I suppose I ought to ask your permission to land here?'

'Whatever do you mean, your Excellency?'

'Well,' replied the Governor, 'they tell me you are known here as the King of Candia!'

thought a German submarine might be going, in the hope that it would run into the nets and detonate the mines. It was a slender hope.

The terrible problem at this time, when of course there was no Asdic and no radio direction-finding equipment, was to locate enemy submarines – an essential preliminary to fighting them! However, some research work had been done at the Admiralty on the hydrophone principle, and one day a couple of specially developed watertight telephones, complete with batteries, were sent out to Captain Stephenson. The idea was extremely simple – you just lowered the telephones over the side and listened for a submarine. Stephenson tried it without delay and the first thing that happened was that the telephone cable was severed by his ship's screw!

However, the principle was a good one and he soon realised that if he lowered two telephones, one on either side of the bows, he might be able to discover the bearing of a submarine in relation to his own ship, which would be a great advance. So he had the telephones connected to a two-way switch on the bridge and listened to them alternately; he had to stop engines, in fact all machinery, including dynamo pumps, to hear anything at all!

All that was needed now was a submarine to practise with – but when Stephenson asked the Admiral for one, he was furious and flatly refused: 'However, after I had pointed out how impossible it was for my men to listen for submarines when they hadn't the faintest idea what they sounded like, he came round to my way of thinking and let me have one'. A school of instruction was then set up – a regular war college in miniature.

Stephenson's main method of attack was to use three motor launches in line abreast, the two outside ones using their 3 'phones to determine the course and direction of a submarine and passing the information to the vessel in the middle, which would make a depth charge attack ahead of the submarine. They had a pretty slim chance of success, either in pinpointing the submarine or in making an effective attack, for, unlike the escort ships of the Second World War, which carried eighty or ninety depth charges, each of those MLs carried only one (even a destroyer had only four). 'What's more, I found that not all my "listeners" could hear properly – some of them were getting corns in their ears through pressing the 'phones against their heads in a desperate attempt to hear *something* – so we established a simple hearing test; would-be "listeners" had to be able to hear a little watch held a short distance away from them.' Gradually the problems were overcome, and experience proved that the hydrophone principle would work in practice, providing the practical method of attacking enemy submarines which was so desperately needed.

Once he had proved that the idea worked, Stephenson was desperately keen to get some more gear, and to check up for himself on the latest developments in this field. He got permission to go home and meet some of the top antisubmarine people at the Admiralty. Then he went all round the country picking up telephone gear and, by the time he had finished, had a whole luggage-van full of the stuff to take back to the Mediterranean, all labelled as his personal luggage:

'Now, as you can imagine, it was a tricky business getting all this lot back across Europe in wartime without interference, and I don't know how I would have done it if I had not brought with me an interpreter, by name Peshandjian (which we changed, by the way, to John Pesham, to avoid awkward questions). On the way back I made him sleep in the van with his toe tied to the door, so that if anyone tried to open it they would wake him up. Twice intruders tried to get in and twice "John Pesham" scared them away. But when we reached Italy, the van was completely disconnected from the train, with Pesham in it! However, he was a good fellow, created no end of a fuss and managed to get a wire to me in Taranto; I raised hell's delight with the Italian Government till they got it back for us. If "Pesham" hadn't been in it, we would never have seen that van or its contents again!'

When the gear eventually did arrive, it was fitted to all the trawlers and drifters, but it's difficult to say how much real success they had; they damaged one or two submarines for sure, but it's doubtful whether any were destroyed. However, these efforts made Captain Stephenson's name as an anti-submarine expert – 'in fact, as far as I know, I was the *only* one!'

The C-in-C Mediterranean, Admiral Sir Arthur Gough-Calthorpe, asked Stephenson how he thought anti-submarine operations could be improved and he replied that, in his view, it was no good trying to win a war by defensive methods alone: all the Navy was trying to do at that time was to protect our convoys, and methods of detection were still so primitive that the so-called 'escort ships' ought, more correctly, to be called 'rescue ships' because about the only useful thing they could do was pick up survivors; even after a submarine had struck, they had no idea where it was nor how to go about attacking it. 'All this is getting us nowhere, sir,' the Captain continued. 'If you've got a swarm of wasps, the best thing is to tackle them in their nest and, if you can't do that, attack them where they are concentrating. Now all the enemy submarines in the Mediterranean have to leave from and return to Trieste, so we ought to have a "barrage" in the Adriatic – lines of ships with hydrophones to force the subs down – lines of observer-balloons to spot them if they try to surface, and to pinpoint them for subsequent attack.'

The C-in-C agreed with all this and got his staff to draft a plan. Then, when it was ready, he told Stephenson to go home and get Admiralty approval for putting it into operation.

'Yes, sir,' he said, 'but when I get to the Admiralty, am I your ambassador, or am I your messenger?'

'I don't understand what you mean, Stephenson. What are you talking about?'

'Well, sir, if the answer to the plan is "yes", then it doesn't matter a damn either way. But if it's "no" and I'm your messenger, I bring back the answer "no". On the other hand, if I'm your ambassador, I do all I can to get that "no" changed into "yes".'

'Oh, I see what you mean now, Stephenson,' said the C-in-C. 'You are my ambassador, of course!'

Stephenson had been right to anticipate difficulties. After he had been about ten days hovering around at the Admiralty, the Naval Secretary to the First Lord sent for him and said: 'Oh, Stephenson, I'm sorry to say that the plan you brought from your Commander-in-Chief is not approved. I have arranged a passage back for you in a week's time, and you can go on leave till then.'

'Excuse me, sir,' said the Captain, 'I'm not going on leave until I've seen the First Sea Lord.'

'You can't see the First Sea Lord!'

This was where 'ambassadorial' rank came in handy.

'You obviously don't realise who you are speaking to,' said 'Puggy' grandly. 'You are speaking to the Commander-in-Chief Mediterranean, and I don't leave until I have seen the First Lord. You can't stop me!'

The Captain saw the First Lord.

The experimental work they had been doing in the Mediterranean, using three launches and one of our own submarines, was explained – how the sub was given a certain amount of time to get away, and then found time and time again with the hydrophones. The system had even been successfully tried out in a destroyer, thus dissipating Captain 'D's fears that stopping the engines suddenly to listen on the 'phones would damage them beyond repair – so destroyers too could be used for anti-submarine work. In fact, they had all the necessary ships and equipment to make a 'barrage' work, and it was a right operation – *attacking* the enemy rather than waiting for him to attack our convoys; this was the way to destroy him.

'At any rate, sir,' 'Puggy' concluded, 'that's what C-in-C Mediterranean says, and I'm sure he'll do it if you only give him a chance.'

The upshot of all this was that their Lordships changed their minds; they decided to set up the barrage and put Captain Stephenson in command of it. It was a pretty big horse for him to ride.

He was Captain (Barrage) – Captain 'B' – till the end of the war in 1918. At his strongest he had 230 ships, which was a lot for a very junior Captain with only a year's seniority, and the organisational problem of keeping eighty of them always at sea was considerable.

A very large proportion of the men were inexperienced, so there was a great deal of training; it was quite like Tobermory, in fact: 'In general, the RNVR officers were for the most part highly intelligent, so it did not take them long to get the hang of what was wanted. However, there was one ship, I remember, which was so bad at drill and so slack that I ordered her to stay out all night on patrol instead of coming into harbour – but I didn't do that again in a hurry. You see, while they were out, they spotted a German submarine on the surface and depth-charged her, and I had to recommend the Captain for a DSO. So much for my punishment!'

There was another ship, too, which ought by rights to have been in trouble but got away with it. This was an American sub-chaser serving under Captain 'B' which mistook a British destroyer for a sub on the surface at night, opened fire and fractured her main steam pipe; the destroyer had to be towed away. Naturally the Captain was very worried indeed and quite astonished when Stephenson commended him for the excellent shooting of his crew: 'I considered his action fully justified, for he had no reason to suppose a British destroyer would be in that particular position. We had American as well as British ships in the fleet, and this created some strain since I knew there was always quarrelling when British and American sailors got together ashore. However, a simple way was found out of this difficulty by making sure the Americans – good fellows though they were – were based a good seven miles away from our people. Thus separated, we got along together splendidly!'

Stephenson had four bases altogether, at Gallipoli, Taranto, Tracasi and Corfu, though he spent his time in his own ship, the *Whitby Abbey* which, in earlier days, used to run butter and milk from Hull to Holland: 'I rather prided myself on not being ashore more than necessary'. Occasionally he had to go to Brindisi, where his senior officer, Commodore First Class Howard Kelly, had four old cruisers, eighty destroyers and eight submarines, and where the Italian Fleet was stationed; they were, of course, on the Allies' side. Stephenson distinctly relished the opening offered him by the Italian Admiral one day when he inquired where he was based. With reasonable honesty he was able to reply at once: 'At sea, sir!'

How successful was the 'Otranto barrage' as it was called? Well, it was not as effective as Stephenson had hoped – and for one very simple reason. They had been experimenting with British submarines and didn't know that the enemy submarines were much quieter – after all, they could hardly get hold of the genuine article to practise on. But subsequent inquiry proved that they had hampered the submarines a good deal; quite a few were damaged and a number had to return to port. This was, at any rate, some measure of success and, to achieve it, the men had to put in a good deal of sea-time to keep the fleet in action by day and night for nearly two years.

This did not always make 'Puggy' the most popular of men, and after a time he discovered that someone had written a scurrilous song about him to the tune of 'Ragtime Cowboy Joe', which was being sung with great gusto in many of the ships. Captain 'B' eventually traced the author – a young Sub-Lieutenant RNVR by the name of Frank Green, a gifted writer – and sent for him. 'I am not fully acquainted with the technique of journalists,' said 'Puggy', 'but I understand that when people write about someone or other they generally give a copy of what they have written to the person concerned. So I should like you, please, to give me a signed copy of your damned doggerel!' Green did as he was asked, and became a lifelong friend.

This is how the 'damned doggerel' went:

### RAGTIME CAPTAIN 'B'

1. In the Adriatic where the MLs are
        And the only light to guide you is the evening star
    The roughest, toughest man by far
        Is Ragtime Captain 'B'.
    He's got a name for shouting like a depth charge boom
    People say his voice is like the trump of doom
    Ev'ry night you'll hear him 'neath the silv'ry moon
        Crooning dreamily.

#### CHORUS

He always shouts like the thunder of a battle
You should see how the MLs all skedaddle
In a way that's half demented
They exhibit such emotion
As they scatter round the ocean:
Off they tear when they hear that fellow swear,
Because they know quite well you see
That the man to make you shake and quake
and keep you night and day awake
Is Ragtime Captain 'B'.

2. When trotting back to harbour with a heart so light
        And visions of a drop of Scotch at last in sight
    'Keep both Divisions out tonight'
        Says Ragtime Captain 'B'.

It ain't no use your saying that your petrol's low
Or, 'Don't you think it's looking like a Sirocco?'
You'll only hear a most sarcastic voice say 'Oh?
    Just go out and see.'

CHORUS

Then once again we resume our damned patrolling
And I think, when a big one comes a-rolling,
'Why on earth was a Captain first invented?'
As I try to read his flashing
And my crockery is smashing,
How divine if he'd only bump a mine;
Oh, I would shout and dance with glee
For the man who's made me lose my views
That MLs were a yachting cruise,
Is Ragtime Captain 'B'.

3. One day a lovely shining rainbow my eyes met.
    Thought I, 'Why that's the loveliest that I've seen yet',
Then with his brow all cold and wet,
    My look-out said to me:
'Those colours that you're gazing at are no such thing,
It's half a hundred thousand flags upon one string.'
I gently fainted, murmuring:
    'Ragtime Captain "B" '.

CHORUS

For his roar half a hundred cables carries
You would think 'twas the gun that's shelling Paris
Or the zoo when they're going to feed the lions.
There is such an awful bawling,
An unholy caterwauling,
But some day a tin fish will come his way,
And I won't fire my gun – not me,
No, I'll heave a sigh of glad relief
And pass the hat to buy a wreath
For Ragtime Captain 'B'.

# 8

# La Belle Petite
## *Dauntless*

(Captain Ashore and Afloat, 1919–1922)

After the end of the war, when the Otranto barrage had been wound up, Captain Stephenson set off for Gibraltar in *Whitby Abbey* and awaited orders to proceed to England. A number of trawlers were also in harbour there, in the same situation – doing nothing but wait for further instructions – a demoralising state of affairs. This, of course, would not do for 'Puggy': 'What they needed, of course, was exercise or work – both difficult to impose on undisciplined fishermen in peacetime. So we had boat races for the different sections of officers and men, and also sent them on marches to the top of the Rock – a prospect we made seem attractive by pretending this was a great privilege specially granted by the Governor.' The trawlers were also allowed out three at a time to catch fish, not with a line but with a depth charge. Each of the trawlers had a boat, and as soon as the charge dropped by one of the ships had exploded there was a mad rush to see who could pick up the most fish before the sharks got them – a piece of unofficial depth-charge drill often practised by convoy escort ships in the Second World War.

At last orders from home were received. *Whitby Abbey* was to take all the trawlers under her charge and, en route for home, call in at Lisbon to clear the approaches of mines.

Shortly after he had begun this work, Captain Stephenson received a cable to proceed home immediately in a sloop. He had been appointed Director of the Anti-Submarine Division at the Admiralty – a very gratifying appointment, as Directors are normally *senior* Captains, but there was no one who could match Stephenson's experience in this field.

Soon 'Puggy' was tackling the problems of Whitehall with characteristic energy. He had a number of scientists working for him at Shandon in Scotland. Their efforts, Stephenson felt, were frequently misdirected: 'They were all very keen – but, as often as not, on their own *individual* objectives. Let us suppose one of them had been asked to find out, say, how to detect a submerged submarine at a distance of 2,000 yards and at a depth of from 40 to 150 feet; in the process he was likely to come upon some new and interesting discovery quite unconnected with his final objective, and would pursue this new trail, forgetting that this was not the job he had been paid to do. I came to the conclusion that to guide and direct brilliant scientists was not for such an ignorant man as I!'

It then struck Stephenson that other Departments of State might have the same problem as he had encountered – finding their scientists' work wasted, or not used, just because it was not carefully directed towards its proper goal. Surely what was needed was a National Research Organisation, to which each Department could have access, while, at the same time, retaining a small separate scientific unit for its own special purposes.

Admiral Wemyss, then First Sea Lord, agreed with these arguments and told Stephenson to get the proposals set out in detail. The best brains of the Anti-Submarine Division were then set to work on a paper dealing with the whole question of research and experiment on a national scale. At this point Admiral Wemyss was relieved as First Sea Lord by Admiral of the Fleet Lord Beatty, who soon expressed himself in cordial agreement with what was being done. He insisted, in fact, that the paper must go to the Cabinet, since it dealt with matters not confined to the Admiralty. Stephenson's whole Division laboured to make the paper as cogent as possible, and on an agreed day it was handed to Beatty, who was to give it to A.J. Balfour, 'by far the most learned and scientific man in the Government of the day. Alas, Lord Beatty forgot to pass the paper on, and the chance was missed!'

As Director of the Anti-Submarine Division, Captain Stephenson was responsible for the development of all antisubmarine devices. It occurred to him that these devices might also he utilised for general navigation and, if accepted as such, that their further development and the training of men to operate them might be paid for by the Board of Trade, thus saving the Admiralty money and providing a pool of skilled men who could later be of use to the Navy in war.

Lord Beatty told Stephenson to go to Lloyd's and to offer the shipping companies three Royal Navy inventions – Directional Wireless, 'Leader Gear' and the Echo Sounder. On the latter, the Captain was a bit shaken when the elderly chairman of P & O said at the luncheon where he made the offer: 'Well, gentlemen, I have always found the Lead and Line very efficient!' An astonishing statement, since of course the lead could be used for taking soundings only in a vessel going at ten knots or less.

'Leader Gear' consisted of a cable laid down the centre-line of a swept channel (that is, a channel cleared of mines). It sent out pulses which could be monitored on board ships, thus guiding them, on one side of the cable, safely through the swept channel. It seemed clear that this had an application in narrow seas like the Channel, for in fact it enabled ships to pass within twenty yards of each other in perfect safety. Indeed, it could probably still be useful; collisions at sea are still far too frequent, in spite of modern navigational aids.

The Admiralty convened a meeting of the Dover-Calais shipping interests, the British and French railway companies and the Channel pilots, under the Chairmanship of Admiral Chatfield, and, at this meeting, acting on the First Sea Lord's instructions, he made a cold offer on behalf of the Admiralty to lay a 'leader' cable from Dover to Calais together with the necessary terminal machinery, completely free of charge – if the railway and steamship companies would agree to maintain the equipment. The pilots immediately saw the advantages of the scheme, but the railway people turned it down flat. Stephenson could hardly believe his ears.

The only one of these three ideas to find rapid commercial exploitation was directional wireless though, of course, the Echo

Sounder was fairly soon developed. It is at least possible that 'Leader Gear', too, could have saved many lives.

Stephenson – with his great interest in the RNVR – served at this time on a committee set up after the war to consider the future of the Royal Naval Reserve and the Royal Naval Volunteer Reserve, and to settle a plan of recruitment and training for their members.

The committee developed detailed proposals for both Reserves, at each stage being careful to get the full agreement of their respective Senior Officers, which tended to ensure that, when they published their report, it would encounter no serious opposition from the Reserves themselves.

There were considerable differences between the RNR and the RNVR: 'The RNR, being professional seamen, were thoroughly disciplined, if a bit blinkered; the RNVR were intelligent, though undisciplined'. But, in spite of their differences as groups and as individuals, they seemed to be thought of as one great mass. The committee recommended that officers should be trained and allocated on appointment for special duties – Gunnery, Torpedo, Navigation, Signals and so on: 'I remember that the Appointments Officer objected to these proposals, as they would make appointments so much more complicated: the chairman of our committee asked him if the Navy should be organised to make his work simpler!'

Towards the end of his time at the Admiralty, Captain Stephenson was much disturbed by the difficulty Navy men found in getting suitable employment on retirement.

Many of his old shipmates called to ask for help in getting a job: 'Alas, the only jobs I could find were as caretakers, for good character was their only qualification. They could find no opening in trade, for they had no vocational training.' Captain Stephenson heard that the Marines had developed such a training scheme, so he got all the information he could and stored it up until he received a sea-going appointment.

His chance came quite soon and, after all the frustrations he had encountered at the Admiralty, he welcomed it. He was appointed in command of HMS *Dauntless*, a light cruiser of the 'D' class, and the new Captain at once put a vocational training scheme into operation – *Dauntless* was the first sea-going ship in the Navy to do so.

The plan was not very well received at first, 'for, in my experience, most men *hate* new ideas'; but attitudes changed when the Captain announced that the training would be held in *Service* time – this made people considerably more eager: 'nor did I have any bad conscience about this for, in my view, when people become enthusiastic about one thing it tends to have a general effect on their lives – certainly I found this to be true in our case'.

The scheme concentrated on two types of work – boot repairing (which was a skill in demand on board anyway) and the driving and servicing of motor cars. Here *Dauntless* was lucky in that a friendly Naval Stores officer let the ship have spare parts to keep on board, and fixed up spells of work in garages while she was in harbour. In this way many became qualified for work that was in demand when they left the Service – and it was not long after *Dauntless* had shown the way, that vocational training was taken up in a big way by the Navy as a whole, particularly in the big shore barracks.

One of the first assignments for *Dauntless* was to go to Avonmouth during a general strike to safeguard the docks and lock gates. Her first contact with shore was a shout through the megaphone as she came alongside: 'Please give us the address of your football secretary!' The next man needed was the vicar: 'I told him we would march our entire ship's company to church on Sunday. I didn't care whether they were Hittites, Jebusites or Buddhists, or whether they worshipped God or Mammon. We made rather a splendid turn-out, complete with our squeegee band!' The vicar addressed the ship's company at the service and thanked them cordially for the manner they had entered into the life of the community and their friendly attitude in general.

There were a good many games of football during that stay. There were concerts, too, and a match with the town's crack tug-of-war team: 'We challenged them and were beaten, even though we had fourteen men pulling for us against their eight. But all this achieved a good feeling between the ship and all sections of the community – indeed, the dockyard mateys went so far as to promise that, if there was any trouble, they personally would see to it that no damage was done!' Eventually the strike petered out and *Dauntless* left Avonmouth after fourteen days without trouble of any kind.

The next piece of work for *Dauntless* came about through an American tragedy. A British airship, the R.38, had been sold to the United States and the American crew were sent over here for trials; during them, the airship fell into the sea and all the crew were drowned. The Government decided it would be proper to send back the bodies of the fifteen Americans in a man-of-war, and *Dauntless* was nominated for the task.

Time was short to prepare for this important voyage and there were several pressing problems to deal with.

First, although this was a mission of great solemnity, Captain Stephenson felt certain that while he was in New York, especially as prohibition was in force, a number of thirsty well-wishers would come on board and would certainly not be satisfied with *one* drink. Stephenson therefore applied for an entertainment allowance of £150. The answer came back expressing blunt disapproval and pointing out that this was not an occasion for conviviality. Now the Captain knew that negotiations were under way between the United States and Britain on the rights of foreign ships in US territorial waters, and that the question of whether visiting ships should be allowed to open their bars was being discussed, with the United States opposed to this: 'So I drafted another letter to the Admiralty saying that I had no wish to hamper these delicate negotiations, and my intention was to close our own bars as soon as we reached territorial waters'. Stephenson got his £150!

The next difficulty was that orders instructed *Dauntless* to return home as soon as the task was accomplished, at economical speed. In reply to this, her Captain drafted a letter on these lines:

'Sir, with reference to my instructions to return home as soon as refuelled, I would like to observe that I have seen on many Royal Navy Recruiting posters the phrase "Join the Navy and See the World". My view of the world in recent years has consisted of four months at Portland, three months at Invergordon, two months at Rosyth and two days at Torquay during the summer, during which there was a south-westerly gale blowing, and there was no contact with shore. I would submit therefore, that, on this opportunity to see a foreign country, my Officers and Ship's Company should have the chance to go ashore at least twice in New York.'

*Dauntless* was given permission to stay in New York for one week.

Thus, the planning problems relating to the ship's stay in the United States were solved, but Captain Stephenson was much preoccupied with his actual arrival there. He was to be met by a squadron of aircraft and would be escorted in by a flotilla of destroyers. The ship was then to proceed to her berth at Brooklyn Harbour beyond Brooklyn Bridge:

'We therefore, quite naturally, checked on the state of the tide at our time of arrival to see whether we could pass safely under the bridge. According to our calculations, based on the measurements we had of our masts, we would have an ample clearance of between three and four feet. But, for some unaccountable reason, I did not feel happy and we sent the carpenter aloft to cheek the height of the topgallant mast; the measurements we had were wrong! We had to lower the wireless mast, or we would, for sure, have hit the bridge – and just imagine what that would have done for the prestige of the Royal Navy – with all eyes on us as we came up the river and all the papers next day full of pictures of HMS *Unready*. And it was simple good luck that saved us!'

The Captain's determination to be fully ready led him to rehearse as many aspects of the ship's task as possible, particularly the tricky business of manipulating heavy coffins on and off the ship. He tried hard to find out the actual weight of the coffins beforehand, but no one could tell him, so he had a mock-up made in Naval Stores weighing six hundred pounds, and the four bearer parties (each consisting of one midshipman with drawn dirk and eight sailors) practised heaving it in and out of a railway wagon and up the gangway to the ship.

The embarkation at Devonport was a great ceremony; a full Guard of Honour from each of the Services and a gathering of Senior Officers in frock-coats and with swords. All went well and, on the other side of the Atlantic, the rendezvous with the escort of six destroyers at New York went according to plan. However, a discordant note was struck when a tug came alongside bearing the Port Commandant, a representative of the US Secretary for the Navy *and* a press party, ready to come on board. The Captain flatly refused to have them cluttering up the ship when he was making a ceremonial entry, but was told he must have them – that, 'if we didn't, they'd make mincemeat of us in the papers. I stuck to my guns – and the press couldn't have been kinder!'

Having safely negotiated Brooklyn Bridge, the next difficulty made itself known when the ship was alongside in Brooklyn Harbour. This was the time of the Troubles in Ireland, and the British Naval Attaché in Washington came on board with a message from the Ambassador, warning that the dockyard workers were mainly drawn from the Irish population of Brooklyn, and that they had laid in a large quantity of bad eggs and rotten tomatoes, destined for *Dauntless*'s liberty men. The Captain was extremely angry and sent a message to the Ambassador saying that the visit should not have been allowed to take place under these circumstances, and threatening to take the ship away at once if there was any trouble: 'then I pointed out to the Superintendent of the dockyard that his and our interests matched, neither of us wanted trouble, and I asked to be moved to the other side of Long Island. As an additional precaution, when our men had all-night shore leave, they were ordered not to return until 9.30 a.m. which, of course, pleased them, and ensured that the workers would be at their jobs in the dockyard, instead of lying in wait at the gates for our liberty men.' There were no incidents during the ship's stay.

*Dauntless*'s time in New York was a great success; there was a ceremonial tour of the city, there were no defaulters among the ship's company throughout the whole time, the ship was inundated with visitors, who did not only drink lemonade (though the Captain came back with some change from his £150), and the officers were dined by the Pilgrims of the United States of America, an occasion where, despite Prohibition, every drink under the sun was available: 'it was, in fact, provided by a good friend of mine – the Commanding Officer of the 36 American sub-chasers which had served with me on the Otranto Barrage; he must have had good friends in the Police Force!'

Later that year, in July 1921, *Dauntless* was called upon to represent the Royal Navy on another ceremonial occasion – the 'Semaine Maritime', a Gala Week at Le Havre. The whole of the French Main Fleet was to be there and an American battleship – and the little *Dauntless* was given just three weeks to prepare. This would have been short enough notice in ordinary circumstances but, in fact, when the Captain received his orders, the ship was in dry dock at Chatham, and half her ship's company was on leave.

Things had to start moving at once. Captain Stephenson asked the British Naval Attaché in Paris if there was to be a regatta; the answer was 'yes', *Dauntless*'s boats were ashore – high and dry in the dockyard – but the Captain managed to borrow two good racing boats from the Chief Constructor. He told the watch on board that they had got to produce two first-rate boat crews in no time at all, and that shore leave would have to be restricted until all preparations were made; they got down to it with great spirit in the evenings and dog watches – and they were a match for anyone by the time the cruiser sailed.

While all this was going on, the Captain contacted the Admiralty and asked that war medals (which had not been generally issued yet) should be issued as a matter of the utmost priority to his ship's company. He went to the Fourth Sea Lord and said he wanted his boats to be decorated with gold leaf and enamel, and he must have a red and white awning fore and aft. 'But you can't have that,' said the Sea Lord, 'it's never done, even for a battleship. And I have no money.' 'Never mind, it won't cost you a penny,' 'Puggy' replied. 'There are some old battleship awnings in the sail-loft in the dockyard and all I want to do is cut them up – they'll never be used again otherwise.'

The Captain got his awnings.

He gave the cook some 7lb biscuit tins and told him to fill them up with fancy cakes and biscuits, and borrowed C-in-C Chatham's band. In every department, his great object was for the ship fully to obey her orders – which were to go one better than the American ship – the battleship *Pittsburgh* wearing an Admiral's flag!

*Dauntless*'s arrival in Le Havre was not very auspicious. She was ordered to anchor near a French battleship but, unfortunately, she was not in sight – nor was the land; the British cruiser's billet was five miles out. After waiting all the forenoon for the customary courtesy of a welcoming call, the Captain decided to get trussed up in his epaulettes to pay his official calls ashore. Unfortunately the British Consul was out of town: 'unfortunately, that is, for both of us, because I had to make another ten mile journey in my small motor-boat the next day, and *he* had to hear some very painful truths about consuls who choose to be out of town when a British squadron arrives (though, to be quite honest, he was a good man and had been misinformed about the date of our arrival).'

The Captain was rather more successful with his call on the Préfet of Le Havre. He had already made up his mind that he ought to be in harbour, instead of five miles out in the Channel: especially if he was going to give one or, perhaps, two big receptions, which was his plan. How to achieve a change of berth? 'Puggy' told Monsieur le Préfet that he wanted to give a party, and that, in particular, he hoped to welcome the Préfet's charming daughter on board as his guest: 'but, alas, we were lying some considerable distance off-shore and travelling all that way in a small boat would surely ruin her pretty dress-the fact was that, in our present position, a party, alas, seemed out of the question.'

The Préfet saw the point at once. 'When would you like to come into harbour?' he asked. 'At 10 a.m. tomorrow, if that is convenient, M. le Préfet. And, since there are apparently two afternoons still available, we should like to give two receptions.' So, of course, that is what they did.

The following morning *Dauntless* and her attendant destroyer came up harbour as planned (the only visiting warships to do so) – getting off their salutes to the French Fleet as they arrived. The first reception that same afternoon went off with great éclat. The First Lieutenant had got the ship looking very fine, the refreshments were good, and the red and white awning looked perfectly splendid. The only unfortunate thing was that almost everyone who came on Saturday came again on Sunday; *Dauntless* had planned for a hundred guests, and had nearer four hundred!

At every stage Captain Stephenson wanted to be sure of coming off best – not for his own sake, but for the sake of the

Navy and the Country. There was a big presidential dinner in Rouen, and Stephenson heard that the American Admiral intended taking his escorting destroyer up the river – 'very well then, so would we. The Americans then changed their minds, but I didn't see why we should. So the Americans changed their minds again. At the dinner itself, the American Admiral spoke in English; thanks to my mother's tuition I was able to reply in French. We did not take part in the Regatta, which was a pity after all the training our crews had done, because the Americans refused to take part; but when I looked at the seating plan for spectators, I found their Admiral had been placed in the front row, and I was to sit behind. I immediately insisted that these arrangements must be changed or no representative of Great Britain would attend.' The British representative sat in the front row.

Such diplomatic quibbles apart, the three weeks of hard preparation, plus great quantities of the Fourth Sea Lord's stores, paid off as far as Captain Stephenson and his ship were concerned. It must have been good to stand by the side of the French President for the review, and to hear him exclaim, as the Presidential yacht came past the British light cruiser with sailors lining the side, smartly at attention, with their arms outstretched – 'Ah! La belle petite *Dauntless*!'

# 9

# Small Tornado

(Captain, 1923–1924)

In the summer of 1923, Captain Stephenson took another step forward in his career when he assumed command of a battleship – he was appointed Flag Captain to Rear-Admiral Howard Kelly, flying his flag in HMS *Revenge* as Second-in-Command of the Atlantic Fleet. A few months before Captain Stephenson arrived on the scene, a certain Lieutenant Lord Louis Mountbatten had joined *Revenge* – and fifty years later Admiral of the Fleet The Earl Mountbatten of Burma remembered his first encounter with 'Puggy' very clearly:

'I had joined the *Revenge* in Constantinople in January 1923 under Captain Merrick. When we got back to England in the early summer, Captain Stephenson was appointed to relieve him. I will never forget his arrival on board like a small tornado bristling with enthusiasm and affecting everybody with whom he came in touch. His first address to the ship's company was a great inspiration and, although *Revenge* was already a happy and efficient ship, she rose to new heights under Captain Stephenson.'

Obviously the young Lieutenant received a wholly favourable impression of his new Commanding Officer; but when Captain Stephenson saw the name Mountbatten in the list of officers who were to serve under him, he did his utmost (before he had actually met Lord Louis) to get rid of him. Here are his reasons in writing:

'I happen to be rather old-fashioned and conservative in my views, and I think all on board a ship should look to the Captain for praise or blame; and I thought someone like Mountbatten, with all his grand connections ashore wouldn't care a tuppenny cuss what I thought of him. So, very naturally, I asked the Admiralty if they would kindly have him removed before I joined. Well, of course, they did not grant my very reasonable request and nobody was more grateful than I that they hadn't, when I discovered what sort of a chap Lord Louis really was.

'Directly I saw the man at work I realised how tremendously intelligent he was, how full of life and vivacity. He had the gift of getting on with people; people, in other words, wanted to do what *he* wanted them to do. And, after all, that's a very valuable gift in anybody, especially an officer.

'He was, in fact, the most successful of all my officers when handling difficult men, though you would hardly have expected him to afford them the time they needed. As soon as we had a seaman drafted to the ship with a rather dingy record, who needed special looking after so as to build him up, I always suggested to the Commander that, other things being equal, he should put him in Mountbatten's division, because, in spite of all the many activities which he had and the calls for him ashore, he would spare the time needed to help the problem boy.

'One of the ways he helped us very much was in running the ship's cinema. He had, of course, spent a good bit of his honeymoon out in Hollywood and had a great many friends out there, so he got all sorts of films for us which had hardly been released for the public at all. One evening, after Admiral Kelly had hauled his flag down and we were once again a private ship, I gave a dinner party for the C-in-C at Devonport and, among others, invited Lord and Lady Louis Mountbatten. We had a film after dinner – and Mountbatten would allow no one else to run the cinema; he just got on with it himself, in his shirtsleeves, and did not dine with us.'

Lieutenant Mountbatten got Captain Stephenson to adjudicate once over a little wardroom problem, while the ship was at Gibraltar. The officers were planning to give a dance and were undecided about whether or not to have champagne – so

Mountbatten came up to see the Captain in his cabin and asked him to go and help them make up their minds.

'Well, now,' said 'Puggy', 'when you are undertaking any action the great thing is to have your object firmly in mind. Now what is the object of this dance? I think you'll agree that it is to give the utmost possible pleasure to our guests, to uphold the prestige of the ship. How is this to be achieved? By seeing that not a single one of your guests lacks a partner for a single dance, throughout the evening. Does champagne make the slightest difference to that? I think not – it will merely add to the cost, and will mean that many of you will spend more than you can afford!'

After this little talk a vote was taken, and it was agreed that there would be no champagne.

That Captain Stephenson was in favour of sobriety for all on board *Revenge* is borne out by another recollection of Lord Mountbatten:

'On one occasion, when he was complaining about liberty men walking about with their caps flat-aback in Plymouth, he made remarks to the ship's company to the following effect: "Although I know that liberty men often wear their caps flat-aback without being drunk, it always gives this impression. What would you think of me if you saw me walking down Fore Street with my bowler hat on the back of my head? You would all think 'the old man's gone on the booze' and you wouldn't be far wrong because I would never wear my hat in that way unless I was drunk. So I want you all to remain sober and wear your caps in such a way as to show the world that you are sober." '

There was a special personal reason why Captain Stephenson came to value so greatly the presence of Lord Louis Mountbatten on board *Revenge*; he had in the fullest measure all the qualities Stephenson had so admired in his father, Prince Louis of Battenburg: 'and many, many years later, I was very much moved when I met Lord Mountbatten again as First Sea Lord; I naturally called him "sir": but he called me "sir", too! That was an extraordinary act of courtesy – there was absolutely no call for it.'

A powerful reminder of those *Revenge* days in the early 1920s came to Sir Gilbert Stephenson in 1969 when he was invited to the commissioning in Liverpool of the last of the Navy's quartet of Polaris nuclear submarines, the latest HMS *Revenge*. He was greatly touched by his welcome, and of course, deeply impressed with what he saw, but in all honesty, he preferred the old *Revenge* of his day.

Captain Stephenson joined the ship at a time when the Fleet Regatta was not far off. One day he watched the No. 1 boat's crew practising, and it was obvious that their pulling left a lot to be desired; but it wouldn't have done much good just to say so – a more subtle approach was needed.

Fortunately *Revenge* had a chaplain who had been a rowing man at university and knew the fundamentals. 'Puggy' asked him to train, quite unobtrusively, a crew of twelve stokers who knew absolutely nothing about pulling. He was to take them well away from the ship for their practices, and allow them to pull anyhow – to encourage the belief that they were useless – when they were within sight: on no account were they to race any other boat until the Captain thought they were good enough.

However, one day this stoker crew went away without the chaplain, before they were really ready for competitive pulling, and they happened to encounter the ship's No. 1 crew, who naturally challenged them. The stoker crew beat the ship's company crew hollow!

After that, everyone wanted to be trained by the chaplain, 'except the crew of the Captain's galley, who allowed me to train them'. HMS *Revenge* made a happy start under Captain Stephenson's command by winning that Regatta.

The ship's company soon came to terms with 'Puggy'. They had to. Nothing missed his eagle eye. But on the other hand, everyone knew they could expect justice and understanding from him too. The Captain introduced a vocational training scheme, as he had in *Dauntless* – and he was much interested in promotion from the lower deck. But what probably counted most was that *Revenge* very soon began to be recognised as the most efficient ship in the Atlantic Fleet, thanks to the drive and enthusiasm which radiated from her daemonic Commanding Officer. L.P. Brindley – who went on to a commission thanks to 'Puggy's' efforts – remembers the extraordinary lengths to which the Captain would go to ensure the supremacy of his ship:

'He went over the side with a chipping hammer, and said to the Commander, "I have now got down to the camouflage paint which was put on in 1917. We must have the ship's side chipped completely!" Officers and Men then did this willingly. The superstructure was also chipped, and grey enamel applied. The ship then shone like a yacht – a wonderful sight, greatly envied by other ships.'

Such spit and polish may seem superfluous by present-day standards – but it generated pride, a sense of belonging, and a sense of achievement. And *Revenge* reaped a flattering reward. In his one year in command, Captain Stephenson had made her such an outstanding ship that Admiral Sir Henry Oliver, on taking over as C-in-C of the Atlantic Fleet, decided to transfer his flag to her. Sir Henry expressed his appreciation in a personal letter: 'I want to thank you very much for all the trouble you have taken about getting *Revenge* ready for me. The result of your work is very much beyond my expectation of what it was possible to achieve in the circumstances.'

And Rear-Admiral Kelly, by this time serving at the Admiralty, wrote a letter of congratulation to the Captain when a team from *Revenge* won the marathon competition (as they had won most others in the Fleet). 'It is', said Admiral Kelly, 'a personal triumph, as I know it is entirely due to you, and I am very glad that your time in the *Revenge* should be brought to an end in a cloud of glory. You mustn't think that your officers and men (and still more your late Admiral) do not realise and appreciate how much they owe to you for the happy and successful time they have enjoyed under your command, and a large part of their pleasure in this success will be that they know it will reflect happily on you.'

# 10

# Controlled Disorder

(Captain and Commodore, 1924–1928)

In August 1924 Captain Stephenson was appointed Chief of Staff to Admiral Sir Sydney Fremantle, C-in-C Portsmouth. This was a very considerable job as, apart from his Service work, the C-in-C was the great man in the town, whose life at that time decidedly revolved around the Navy.

One event during Captain Stephenson's appointment was the Services and Town Fête and Fair on Southsea Common – a big affair in aid of service and town charities on the lines of the Royal Tournament.

The Chief of Staff was Vice-Chairman of the show, and was quite determined the civilians should do their bit. He and the committee absolutely refused to make any move until some of the wealthy townsmen had put up a guarantee of £500 – 'for if we had had war, or severe gales, a lot of money could have been lost. As it turned out, we made a good profit: we gave two shows a day for five days and each performance drew large crowds.'

Another job 'Puggy' inherited as Chief of Staff was that of Chairman of the local branch of the National Association for the Employment of ex-Regulars. This was right up his street, well in line with his work on vocational training. However, this was a time of national unemployment, and if one man got a job it probably meant that someone else lost it; so although there was no doubt in the Captain's mind that ex-servicemen deserved to get suitable employment, there still lurked a feeling that work should not be obtained at the expense of a fellow-countryman.

And so 'Puggy' became gripped with the idea of migration to the Colonies and Dominions. He was in touch with the Empire Migration Office and became a member of the Empire Migration Committee. In the course of this work he made two great friends – Jack Joseph, a director of Lyons, and Commissioner Lamb of the Salvation Army.

Lamb concentrated mainly on sending boys overseas, and 'Puggy' enjoyed pulling his leg about this.

One day he said to Lamb: 'You send many boys out to farms in Canada, don't you?'

'Yes,' he said, 'we do.'

'Are you a moral man, Lamb?'

'Yes, certainly – why ask?'

'Well, when your boys on farms in the country grow up and feel the need of female friends, what happens? They will either drift away to the towns or go with men! Do you call that sound?'

Unfortunately none of the organisations 'Puggy' consulted were keen to send out single women: so he conceived an idea which he put to Jack Joseph:

'Jack,' he said, 'I want you to start branches of Lyons in Canada and Australia and staff them with your young women. Within three months out there, they will get married and you can send out replacements!'

Joseph was all for it and was making plans, when suddenly – he died. And so of course did the scheme.

It appeared to 'Puggy' that emigration was needed for civilians as well as ex-Servicemen, so together with the Lord Mayor of Portsmouth, he formed a local migration committee and got L.S. Amery, then Minister for the Dominions, to speak at the opening meeting.

The committee really wanted to send out whole families, but no such scheme existed. One day Captain Stephenson heard about a plan for Western Australia: the idea was to form complete family communities with a farmer or two, builders, carpenters, plumbers, doctors and so on. This seemed to be just the thing. The Army had an organisation at Catterick where family groups could be taught to build and equip a house and manage poultry; so up to Yorkshire went families from 'Pompey' to train in house-building. Two of them actually went out to

Australia, but they came back pretty quickly. The land they had been allocated was dense forest and it would have taken years to clear!

After his two years as Chief of Staff, 'Puggy' was appointed Commodore of the Royal Naval Barracks at Portsmouth. This, he felt, was impossibly large for a single command and far too impersonal, for there were between 5,000 and 7,000 men in the barracks at any one time: a better idea would be to run the place like a squadron of ships and to give Divisional Officers the powers and responsibilities of a Commanding Officer at sea – indeed he proposed this to those in power, but without avail.

It was very difficult for a Commodore to maintain any real contact with the men, so 'Puggy' decided he would take a regular forenoon stroll around the barracks: 'I simply walked round with a messenger, and everyone knew he could just come up and talk to me about anything that was on his mind. It was important that the men should have a safety valve for any grievances, and I encouraged Divisional Officers to give ratings full opportunities to protest – but some of them thought this dangerous.'

There was one spot of bother when it came to the Commodore's notice that the accounts of the Petty Officers' Mess were being fiddled, and on investigation this proved to be true. So 'Puggy' told them, with regret, that as they could not run their affairs more competently, he proposed to take their accounts for the time being into his own hands – an action interpreted in some quarters as perhaps a bit highhanded.

Indeed, at about this time, the C-in-C told the Commodore he had had a letter from the Admiralty suggesting that the barracks were in a state of near mutiny! 'Puggy' knew this to be fabricated rubbish, the work of some unscrupulous troublemaker. However, the C-in-C, wishing to investigate further, said he wanted to speak to the PO's mess without the Commodore's presence – a proceeding which in 'Puggy's' view, could simply not be allowed. Eventually he found a way round the problem; he made it clear in advance that the C-in-C was to appear by his express invitation.

There was an instance of the essentially good atmosphere of the barracks when the Prince of Wales came to inspect them.

The Commodore asked him if he would like to see some of the new entries exercising in the gymnasium. He agreed – and, as they got near the gym, they could hear the most colossal uproar.

'Where are you taking me?' asked the Prince, 'The monkey house?'

However, as they entered the building, a whistle was blown and there was complete silence, only broken after permission had been given to 'carry on'. The noise, according to 'Puggy' was not in any way a sign of anarchy: 'it was what I call "controlled disorder"!'

Some of the Commodore's methods of establishing good feeling were more suggestive of a holiday camp than an RN barracks. Taking the view that men who sing cannot be unhappy, he instituted the practice of community singing at boxing meetings in the barracks, before the fights began. This was a great success which received much press publicity and, in fact, was the direct inspiration of the great *Daily Express* Community Singing Campaign. The Commodore even got sailors singing on the march through the town, though the gunnery officer in charge of ceremonial drill was very shocked when he first suggested it.

It was clear that, by and large, the men in the barracks were reasonably contented. The idea that mutiny was afoot was most unlikely in an establishment which won almost every athletic contest going; they even won at football, which was unheard of – for Whale Island, the gunnery establishment, had always won that for as long as anyone could remember.

The regatta presented perhaps the greatest challenge of all, for the barracks at Portsmouth are some distance from the water. However, the Commodore, with typical ingenuity, got the engineer to design a land boat, with a device to put a resistance on the oars roughly equivalent to that of water: 'and this enabled us to put in more practice than we could possibly have done otherwise. The result was most gratifying. We won every event!'

Yes, some of the methods which helped to create good morale in the barracks were decidedly unconventional. For example, the stokers' mess was badly in need of redecoration, but the dockyard people made difficulties over it. So the stokers were asked if they would like to do it themselves, as the dockyard seemed unwilling or unable to carry out the work. The stokers responded to this challenge by redecorating the entire mess in a single night – starting as soon as the dockyard workers had gone home and finishing before they returned the next day. The stokers were delighted – not least, perhaps, because their achievement was, strictly speaking, against orders.

It was most gratifying to find the barracks gaining a good reputation on all sides, and a great honour for 'Puggy' when he was asked to command the King's Birthday Parade of June 1927, a task normally reserved for the gunners of Whale Island. There were some 7,000 men taking part from all three services, so it was a really big affair, and it went off very well. But it might easily have been most unfortunate as far as the Commodore was concerned. As he was getting into his car after the parade was over, his sword-belt came unstitched: 'just imagine how it would have looked if it had happened during the ceremony – for I stood right at the centre of it all, giving the orders!'

'Puggy' was now at a critical point in his career. Was he to go on to flag rank, or to be retired when his time in the barracks came to an end? Whatever the answer was to be, he wanted to know it, so in the summer of 1928 he went to see the secretary to the First Sea Lord and asked what his future was to be, as he had many arrangements to make if his time was coming to an end: 'He told me I was *not* going on. For the next six months I never told a living soul, not even my wife – with the result that, at my farewell concert in the gym at the barracks on 3 July 1928, and at my farewell dinner, people were still giving me good wishes for a flag appointment.'

Early in 1929 the Commodore was promoted Rear-Admiral and the day afterwards placed on the Retired List. His first career in the Royal Navy had come to an end.

# Civilian Interlude
## (1928–1939)

# 11

# Monkey Whiskers

(Vice-Admiral, Retired, 1928–1939)

A period of eleven years at home now lay ahead for the Admiral. So far we have said very little of his life outside the Service – indeed, a reader having proceeded this far may almost have assumed that he had no family. Far from it; his family had occupied a large place in his life.

The Admiral was intensely proud of his wife. 'She was not only most attractive and friendly – she was very wise. It was a matter of great satisfaction to me to see how she was welcomed by all my shipmates and how they would talk to her on matters which they could not discuss with me. I don't think I ever returned home without finding her there ready to welcome me. She died when we had been married a bit over fifty years – and they were fifty years of great happiness.'

The Stephensons had three children – Nancy, Robin and Gilbert. None of them is now (1999) alive. But in 1970, Sir Gilbert told me about them all with great pride: 'Nancy possessed a strong character and was, I believe, as much a problem to her mistresses as she was to her parents. She was a pupil at, I think, eleven schools – this not by reason of her expulsion, but through my changes of appointment, though at times the threat was there and one hoped there would be a change of appointment before it could occur! She was quite fearless, and excellent company – she always had many friends; her governesses were not of this company. My wife and I had always expected her to marry. She did not, however, and offered the explanation that those she would have liked to marry were already fixed up, and that those who would have liked to marry her were not her cup of tea.

'A profession, of course, was necessary, whether she married or no. She herself had no preference at all, so we decided she should train as a children's nurse. Accordingly she did a course for this work. She disliked it so much that, although she completed the course, she did not earn a certificate. Think again! She chose art – painting – and went to the Technical College at Portsmouth. She painted well but not, in my opinion, well enough to keep her in comfort. Think yet again! This time it was dressmaking – a craft for which she showed great skill. But she had no ideas of finance and, as far as I gathered, worked for those who either could not afford to pay or did not trouble to pay. She lived at home, so money was not a pressing need.

'When my wife died, I asked her if she would like to run my house. She agreed gladly – and how well she does it! And this is but a small item in her life. She goes out speaking at least three times a week, is chairman of a flower club, helps with meals on wheels and visits scores of lonely people.

'My elder son Robin chose the Army. I must admit I hoped both my sons would choose the Navy, but when I returned after the First World War I found Robin had joined up with the soldiers training on Wimbledon Common; he used to come home for meals perched on the Colonel's horse. There was no doubt that he was for the Army, and I never suggested any alternative; but he is a great boat-sailor and most skilful with bends and hitches. He joined the tank regiment and, after a short spell in North Africa, he was detailed for training landing craft crews. So well did he do it, that his many applications to rejoin his regiment in the fighting and thus find promotion were refused. He suffered for his skill, and I have always admired him for not complaining.

'Neither did my younger son, Gilbert, follow me into the Service. And this was a disappointment. After five years at the Imperial Service College at Windsor he entered Pembroke College, Cambridge. There he took a full part in the life of the

*c.* 1888: Schoolboy Stephenson – already a sailor!

1897, aged 19: Kitted out for the Benin Expedition.

1916, aged 38: Commander, Crete Patrols.

Captain Stephenson, Chief of Staff to Commander-in-Chief Portsmouth, Admiral Sir Sydney Freemantle.

1927–8: Commodore RN Barracks, Portsmouth.

c. 1942: Commodore Stephenson, RNR, welcomes the First Lord of the Admiralty, Mr A.V. Alexander, with his naval secretary, Rear-Admiral Frederick Dalrymple-Hamilton, on board HMS *Western Isles*.

November 1969, aged 91: The 'Terror' takes his first great-grandson, Richard Bartlett, in hand.

HMS *Active*. Gilbert Stephenson served in her as Midshipman of the Main Top.

Crest from the German Vice-Consulate at
Chios – one of the spoils of a piratical raid, 1915.

Captain Stephenson's two sea-going commands in the 1920s. HMS *Dauntless* (top) and HMS *Revenge* – from watercolours by Donald Maxwell.

1916: In the Kaiser's Navy. Cdr Stephenson, RN, disguised as a German soldier.

A family at war, March 1940. Left to right: Captain Robin Stephenson, Royal Tank Regt; Nancy Stephenson, Red Cross Nurse; Commodore and Mrs Stephenson; District Officer Gilbert Stephenson.

In Tobermory: HMS *Western Isles* on Christmas Day. No chance to relax after the festive meal – there were pulling races round the harbour!

Tobermory: The naval landing stage was in the foreground.

March 17, 1944: Some of the corvettes and frigates at anchor in Tobermory Bay.

1943: Commodore Western Isles with the US Commander-in-Chief European Waters, Admiral Stark and the British Vice-Chief of Naval stuff, Vice-Admiral Geoffrey Blake.

The *Western Isles* crest. Motto: 'Let 'em learn!'

The 'Terror' as Jack saw him during the Tobermory days. Cartoon drawn by a Canadian A.B. of HMCS *Forest Hill* and sent to the Commodore by the ships' Captain as a parting shot!

The Price of Peace. Opening a typical English fête on a typical summer's day.

1955: The Hon. Commodore Sea Cadet Corps is received on board.

1963: With Sir Horace Law, then Flag Officer, Sea Training, at Portland. 'That may be all right for you, but it would never have done at Tobermory!'

HRH Prince Philip at a Western Approaches Command Reunion Dinner, with the President, Sir Gilbert Stephenson, and Chairman, Capt. R. W. Ravenhill, CBE, DSO, RN.

1968: HMS *Eaglet*, Liverpool. A reminder of the gretaest Escort Commander of the Battle of the Atlantic – Capt. F.J. Walker, RN. After a Battle of the Atlantic Service in Liverpool Cathedral, Admiral of the Fleet Lord Mountbatten invited Vice-Admiral Sir Gilbert Stephenson and six memebers of Capt. Walker's Old Boys' Association, togther with their President, Mrs Walker, to be photographed with him – a typically generous gesture to the wife of an heroic sailor.

1960s: Dining with the Croydon Unit of RNVSR – on the Admiral's left Sir John Lang, former Secretary to the board of the Admiralty.

1970s: Annual dinner of the RN and RNVR at Chatham Barracks. Sir Gilbert Stephenson flanked by Rear Admiral Godfrey Place, VC, RN, and Lt-Cdr John Owen, OBE, RNVR. Right: May 1971: Sir Gilbert Stephenson about to read the Naval Prayer at the Battle of the Atlantic Memorial service in Liverpool Cathedral.

1970s: Sir Gilbert Stephenson appears on the Anglia TV programme 'Reflections'.

Liverpool, summer 1971, aged 93: Sir Gilbert Stephenson cuts the tape (made of naval cap-tallies) and officially opens the new headquarters of the Royal Naval Reserve Mersey Division, HMS *Eaglet*.

'And then what did you tell the Admiralty?' Sir Gilbert Stephenson and the author working together on this book. (Credit: Mark Gerson FIIP/ARPS)

August 1971: Sir Gilbert Stephenson and his daughter Nancy at their home at Saffron Walden, Essex. Over page: 'To tell the *truth*? Do you mean to say you don't always tell the truth?' (Credit: Mark Gerson FIIP/ARPS)

College and became Captain of the University Boxing team, winning his fight against his opposite number at Oxford.

'From Cambridge he entered the Colonial Administrative Service and was posted to Northern Nigeria where he undertook in turn the whole gamut of jobs from that of a junior administrative officer to a resident in charge of a province. He finished up as Secretary to the Commissioner for Northern Nigeria in London, a Nigerian, and for a time acted as Commissioner.

'On return to this country he took a leading part in assisting in the establishment and running of Voluntary Service Overseas (VSO) and was the national secretary of this organisation. After ten years with this excellent scheme, which he saw grow from tiny beginnings until it had over 1,500 volunteers working abroad at any one time, he left it to take on a bursarial post at his old College, Pembroke. There I am sure his administrative experience and undoubted ability to deal with young people will be an asset to the college.

'He has a charming wife whom he met in Nigeria, and two delightful daughters. Thank God he lives only four miles away from my home – within borrowing distance, as he calls it.'

To return to the main character in our story: around the time of the Admiral's retirement in 1928, he was very much engaged in work with young people.

For two years he had taken a boys' camp for a fortnight on Hayling Island – this was for an organisation called the Portsmouth Brotherhood. Their great event was a big meeting they had every Sunday afternoon in a large Baptist Hall, with singing and a guest speaker. One afternoon Lord Jellicoe was there, and 'Puggy's' own C-in-C, but the people were shouting for him to speak. 'Look here,' he said, 'junior officers dare not open their mouths, let alone speak, in the presence of their superiors. With Lord Jellicoe here and the C-in-C, it's lucky I'm alive!'

When he retired, Admiral Stephenson was asked to be President of the Portsmouth Boy Scouts, but at first he refused because he felt he didn't know the first thing about the boy scouts. But when they pressed him, he said: 'Right, if you're really in a hole, I'll go through a course and train'. He did a month's training and got the wood badge – the qualification for scouters. While he was under training, a boy addressed him as 'Admiral' and he got told off – no titles were to be used. 'Oh, I'm sorry,' said the boy, 'I thought it was his nickname; I didn't know he was an Admiral!'

After he'd done his training he began to train scouters but, although they chose the days, times and place of training, they could not, or would not, turn up punctually, and that wasn't good enough for 'Puggy', so he gave it up.

He was living fifteen miles out of Portsmouth in the Hampshire village of Shedfield. Here he made up his mind once again to do something about the problem of unemployment, no less pressing in the country than in the industrial towns.

He noticed that the local woods needed thinning and that the cottages were in need of fuel – two facts which suggested a scheme to his mind; so a meeting of unemployed men was called, and the Admiral laid his plan before them. The men were to work together to chop down unwanted timber, saw it up and carry it in lorries to the cottages which needed it. The Admiral was to decide which trees were to be cut down, give instructions about felling, allocate the fuel as required and provide the tools. However, the scheme was short-lived. As soon as a man had seen the first load delivered to his own cottage he would say: 'Well, that's my lot. I'm off!' – and all the others followed suit.

Rather more successful was the boys' club 'Puggy' formed at Waltham Chase, a mile or so away from Shedfield. This was, in the Admiral's phrase, 'a settlement of rascals – a slum in the midst of the country'. The plan was to keep the lads busy all the time at the weekly club meetings; they used to start with a couple of hymns, then have boxing, matchstick drawing and acting. They were totally undisciplined, but that problem was solved in brisk naval fashion with the help of a short length of rope applied three times to the stern, or dismissal from the club. This didn't put the boys off apparently, for they would hang around with their noses glued to the windows, waiting to be let back in.

In spite of these tough tactics, the people of the district still recall the Admiral with affection and respect. And by this time he had acquired a new nickname – 'Monkey Brand', because his bewhiskered visage bore a resemblance to the monkey on tins of a well-known household cleanser. A.J. Lynch is one of many many people, in whom 'Monkey Brand' has taken a lifelong interest. He was ten years old when Sir Gilbert first went to live at Shedfield, and remembered his Boys' Club gratefully:

'All manner of games were taught, but in particular boxing. "Old MB" taught this personally, being a great believer in the straight left – "keep leading with your left-keep your guard up-tuck that blasted chin into your shoulder or you'll receive *this"* – *(this* was a straight left which rocked the pupil right back on his heels).

'Sometimes "MB" would grab five or six boys and direct them to enact any particular historical event learned at school. The utter tomfoolery of it all always reduced any adults present to tears of laughter.

'Portsmouth being only thirteen miles away, "MB" laid on trips to the *Victory* and any RN ship moored alongside . . . someone in the ship's company always had to provide tea and buns for 20–30 ravenous youngsters. In fact, it became reasonably obvious that "MB" was recruiting his own fleet for the RN. "If any of you blasted scallywags join the Army I shall have great pleasure in wringing your dirty scrawny necks." '

Like several other boys, A.J. Lynch was accepted by the Marine Society for training on the *Warspite* and his subsequent history indicates how tenaciously Sir Gilbert 'followed up' his early training and continued to watch over those he had taken under his wing.

At fourteen years of age, Lynch had never been to London before, so the Admiral went to see Mrs Lynch and told her: 'The little so-and-so will be met at Waterloo by Tubby Clayton (the founder of Toc H). He will be fed, watered, and prepared for his entrance to the Marine Society in the morning. There are no expenses involved.' And the Admiral's last warning to Lynch was: 'Don't ever try to pass through London without calling me – do I make myself clear?' This caused Lynch to have a number of interesting lunches and teas in the city: 'We always walked everywhere, and I had great difficulty in keeping up with him. His walking speed was never less than seven miles an hour. "Damned if I know how you won your weight in the boxing contests," he would growl, causing the odd waiter to jump back several feet.'

Because of a mistake in the eyesight test, Lynch failed for the Royal Navy and subsequently served in the *Orontes*. At the end of each trip he had to report to the Navy League and describe the voyage over a meal. Then: 'Did you write each week to your mother? Does she know your time of arrival home? Why not, blast you?'

Eventually, along with many other youngsters, Lynch was put 'on the beach' at the time of the shipping slump. When he informed 'MB' of this personal catastrophe, his fist almost shattered the office table. He then asked if Lynch would consider any other form of employment – and, in the hope of arriving home in one piece, he agreed.

Only a few days elapsed and he received a telegram: 'Report to me Navy League 18:00 hours 18th fully kitted' signed 'Stephenson'. He had arranged for Lynch to join the staff of the Countess of Strathmore.

Many years later, when Lynch was Officer Commanding RAF Air Sea Rescue at Bridlington, East Yorkshire, he attended the opening ceremony for a new Sea Cadet HQ in the town. He was, to say the least, shattered when he learnt that the inspecting officer was none other than 'MB' himself – then at least 80 years young: 'He marched up to me, glared balefully at my No. 1 RAF blue and exclaimed: "Lynch, what the hell are you doing in that rig? How are you? Still leading with that straight left? Put on a lot of weight, haven't you? See you after the parade!" '

Back in the late twenties, 'Monkey Brand' was well settled at Shedfield: apart from the boys' club, he was president of Toc H, and was on the church council. He was not ambitious for money or an important job. But one day Sir Roger Keyes, who was then C-in-C Portsmouth, invited him to lunch to meet Lord Lloyd.

After lunch, Lord Lloyd drew the Admiral into the bow window and said: 'I'm President of the Navy League and our Secretary is leaving us. I think you'd rather like the job. There's not a lot of money, I'm afraid – the salary is £600 per year.'

'Oh, that's very kind of you,' said the Admiral. 'I'm very interested in the Navy of course, but I'm very busy and I've made arrangements for settling down now. I've got enough money – you'd better get someone younger than I, someone who needs the salary.'

Lord Lloyd however was the sort of man who would not accept opposition, and somehow he made the Admiral take the job on – but, not surprisingly, on his own terms. He would only be in the office when he wanted to be there, he would never be there on a Saturday, he would run it exactly on the lines he thought right! On these conditions he became Secretary of the Navy League, a job which included running the Sea Cadet Corps.

The Admiral soon warmed to the work and carried it out for three years with wholehearted enthusiasm:

'It was the time the Navy was great, and we had big meetings in support of it all over the country. The Navy was *the* Service then, not what it is today-tucked away.'

Then in 1935, 'Puggy' got anaemia and very nearly died, and, although Lord Lloyd said he could have six months' leave, he insisted the job should go to a younger man. He was certain war was on the way and wanted them to have someone they could really rely on.

Meanwhile, in the early thirties the Stephensons had moved to a fairly big house, Springwell Place, a short distance from Saffron Walden in Essex. Here the Admiral farmed about seven acres – mostly orchard with apples, pears and plums – and grew a few vegetables.

Springwell is in the parish of Great Chesterford, and Rupert Doble, the vicar's son, at that time a fifteen-year-old lad, 'as much in awe of the Admiral's bluff commands and monkey whiskers as any boy seaman', remembered the Admiral ashore vividly:

'The Admiral, of course, read the lessons at the afternoon service at Little Chesterford church. It is probably true that, if he didn't like the prescribed lesson, he read something of his own choice. I doubt whether the Admiral saw eye to eye with my father on politics but this did not prevent him from taking a very active hand in parish matters.

'I particularly remember a summer garden fête in the vicarage garden, for which occasion the Admiral arranged a traverse – not quite a bo'sun's chair, there were only handles to hang on to – over about 100 yards, starting about thirty feet above the deck from a massive sycamore tree and descending to a few feet above ground level. The erection of the cable was treated properly as a naval evolution, accomplished unhesitatingly by farm lads, with the inevitable fat boy to act as anchor man as we snatched the cable to get it taut. We all knew what "handsomely" meant after that morning. Unfortunately King's Regulations and Admiralty Instructions did not extend to the village lads, who were supposed to pay their sixpences at the fête to make the traverse; most were distinctly timorous at launching themselves into thin air some great height above mother earth. I don't remember the Admiral giving a demonstration run, either, but he solved the problem by persuading a pretty 17-year-old lass to have a go at his expense – after that the lads needed no further encouragement. One even broke his collar bone by trying to demonstrate that it could be done single-handed!

'The Admiral gave tennis parties at Springwell. The grass court was adequate enough, the wire fence was virtually non-existent, and there was a great deal of rough around the court. The Admiral solved this problem by inviting not less than a score of players and providing sixteen of them with billhooks, scythes and sickles and inviting them to get on with it. I fancy the rough extended to the very boundaries of his property. We were always rewarded by a good tea – our elders probably by something better.

'This was when I discovered that he had a music-box cum toilet-roll holder which played a delightful Austrian ditty if you helped yourself to a ream or two of tissue.'

During these pre-war years at Springwell the Admiral did a great deal of speaking in support of R.A. Butler, who was Member of Parliament for Saffron Walden. He used to talk on foreign policy and would put in one or two stories to liven things up. At the end he always called for questions on these serious matters, and on one occasion a woman put up her hand. 'Yes?' said 'Puggy'. 'What is your question, madam?'

'Please, sir, won't you tell us another story?'

Butler and the Admiral differed completely in their views on foreign affairs; Butler didn't seem to mind, but it was probably rather confusing for their audiences.

After one big meeting they had a few months before the war, the gardener's wife came to Mrs Stephenson and said 'Look here, ma'am, we don't know where we are. The Admiral comes and talks to us in the Town Hall and tells us all to dig, dig, dig for food. He says we're going to have a big war and we've got to get every bit of land cultivated that we can. Then along comes Mrs Butler and says: "You'll be glad to hear I can tell you from my husband that there's not going to be any war". What are we to believe?'

The answer was not long in coming.

# PART II

Stephenson, RNR
(1939–1945)

# 12

# This Officer, Without Cap . . .

## (Commodore, RNR, 1939–1940)

When trouble came, it did not take the Admiral long to get back into uniform.

When war broke out in September 1939 he was appointed a Commodore RNR – the rank for the job he was to do as a Commodore in charge of Convoys. In fact, he only had about four months in this work, and they were relatively peaceful; it was before intensive air attack and wolf pack tactics had been developed by the enemy, and his convoys lost only one ship.

However, during this short period the Commodore travelled a good many thousand miles in very varied conditions. On his last trip – to Halifax, Nova Scotia – the rigging was thick with snow and ice and the weather was very foul: the trip immediately before that, he had gone to Freetown, where he spent Christmas baking in the run and bathing in the warm sea. In fact he suggested to the Admiralty that if they wanted to get rid of their aged officers they might choose a more humane method than the hot-cold bath process!

If threats from the enemy were few and far between, there were of course the usual hazards of the sea to cope with: fog is always one of the most unpleasant.

On one occasion the Commodore had reached America and his convoy had been dispersed to various different ports. His own ship was waiting to go into harbour, but the fog was so thick that the skipper wouldn't risk it: he insisted on remaining outside at anchor, waiting for the fog to clear. Now 'Puggy' was very anxious to get in, and, remembering his skipper was a Scotsman, said to him: 'I'll bet you half-a-crown we don't get in tomorrow.' 'Taken!' said the skipper. In they went. 'Puggy' lost his half-crown but it saved him goodness knows how many more days at sea.

On his way back from the States there was very thick fog again; the convoy had to round the south coast of Ireland without any opportunity to fix its position, so they simply didn't know where they were as they made their approach in line ahead, by night, to Liverpool.

Although it was quite against orders, the Commodore ordered full navigation lights to be switched on for, in his opinion, the danger of collision was far greater than any threat from the enemy.

He was on the bridge with the skipper of his ship and his leading signalman, Dolley, beside him when suddenly they saw lights – green, red; green, red; green, red – to right and left and ahead of them. Dolley exclaimed: 'Is that another convoy, sir?' 'No,' 'Puggy' replied, in a state of considerable tension, 'it's a girls' school out for a walk!'

Just as well the ships of his convoy had their lights on, and that they were in line ahead. If they'd been blacked out and in columns, there would surely have been many collisions. As it was, 'Puggy' was glad to see the other ships, from one point of view. He signalled to their Commodore 'Please give me your position' – and at last they knew where they were! They were level with Liverpool, just right to turn in.

One of the most pleasant things about this period of 'Puggy's' life was the way anybody on board ship who had been in the Service – '*the* Service!' – used to come up to his cabin to talk about old times.

Leading Signalman Dolley (a fireman in civilian life) in his position on the bridge, had more opportunity than most for such chats.

The Commodore would inquire about his home, garden, family and hopes for the future in the most friendly way; and when he left convoy work, Sir Gilbert wrote to Dolley recalling their 'happy times together'. He ended: 'I look forward to seeing you in all your glory as a superintendent after the war at a fire brigade display'.

Early in 1940, when he was 62 years of age, the Commodore was recalled to the Admiralty where, at one point, he had no less than three jobs and three staffs working for him; they used to come along from time to time and say: 'Sir, we're rather tired!'

The first of his tasks was to set up and run anti-submarine patrols off the coast of Norway. It was a big job and presented many problems, not least the question of finding suitable people to man the vessels.

One day 'Puggy' thought he would have a bit of fun. He went to the head of 'M' (Military) Branch and put one or two little questions to him:

'First of all, who signs the commissions for these people I am to employ? – Does the King do it? Do you? – or do I? Secondly, what sort of pensions am I to offer their families if they get killed? What flag am I to fly? Oh – and another little detail; I shall need £60,000 in cash, preferably in gold, to carry out a few of my duties. Now what do you advise me to do about these matters?'

'Oh,' replied the 'M' Branch man, 'I think you'll find things will sort out very well when you get there!'

'Thank you very much,' said the Commodore, 'I appreciate that. You, with all the vast resources of the Admiralty behind you, can't answer *one* of my questions; but I, an ignorant little Naval officer out in the wilds, can find an answer to them all. Thank you for the compliment you are paying me. Good morning!'

Another problem 'Puggy' put to the Admiralty was his need of a ship in which to carry out his duties off Norway. 'Oh,' they said, 'you'd better go and get one!' So he went round the clubs in London asking if anyone had seen a ship that might suit him. Finally someone said they knew of one working off Portsmouth, picking up practice torpedoes. This ship was *Philante* – Tom Sopwith's motor-yacht; she would be ideal. So the Commodore went to the Fifth Sea Lord, who at once sent off a wire giving orders that *Philante* be recalled and placed under his orders.

The next job was to start collecting gear and people to man her. Communications would be vital, so 'Puggy' combed the Post

Office for suitable telegraphists. No hope – the Army had got them all; however, he managed to get some back. Then he desperately needed someone who could speak Norwegian, German and English – not easy to come by. However, the Admiralty's Personnel Department managed to produce such a man very quickly.

'How soon can I have him?' asked the Commodore.

'Well, he knows nothing whatever about the sea, so we'd better give him some preliminary training. You can have him in six weeks.'

'No use at all. I want him within twenty-four hours.'

The Admiralty let 'Puggy' have the man: 'I took him on board *Philante* and told him he could do his basic training – *after* the war!'

In fact, the Commodore never got to Norway, for that country fell to the Germans in April 1940. This was when his second job at the Admiralty began. He was to organise those Norwegians who managed to escape to Britain and allocate them to our fighting Services. Difficult – for the Free Norwegian forces could not make up their minds which men they wanted to keep themselves.

Meanwhile 'Puggy's' third and ultimately most important job had begun.

The First Sea Lord, Admiral Sir Dudley Pound, had told the Commodore he wished him to create a working-up base for escort vessels, both British and French, at Quiberon Bay on the French coast near Lorient. Not knowing many people in the Service then, and in order to have at least one person with him whom he knew and in whom he had confidence, 'Puggy' asked Admiral Sir Dudley North to release Lieut.-Cdr. R.H. Palmer, RNVR, who was then serving under him at Gibraltar, and who had been with the Otranto Barrage in the First World War. Unfortunately the fortunes of war made Lorient a German and not a British base, so that plan was abandoned.

However, escort vessels very badly needed training, so a suitable base was sought in the British Isles. Tobermory, a little town on an almost landlocked bay in the Isle of Mull on the west coast of Scotland, was chosen, and 'Puggy' set off around the country looking for a ship suitable for use as a Headquarters. He found the one he wanted laid up in the Kyles of Bute in April

1940. She was a small passenger vessel which had been on the run between Liverpool and the Isle of Man, and she was handed over to Barclay Curle on the Clyde to be completely gutted, except for the boilers and engines, and refitted. She was to become the first HMS *Western Isles* and her conversion was supervised (as were the alterations to her successor) by Lieut.-Cdr. Palmer.

While *Western Isles* was being got ready, the war in Europe had reached a desperate stage. Towards the end of May 'Puggy' was ordered to go to Dunkirk to assist in the evacuation of the British Expeditionary Force. His job was to take charge of the eastern end of the evacuation area off La Panne.

As soon as he got there, on the morning of May 30, he made it his business to make contact with the Commander-in-Chief, Lord Gort. Conditions were chaotic and he had to wade ashore, up to his waist in water. On the beach he met three officers, one of whom immediately left the group to speak to some soldiers who had come without their rifles. Said 'Puggy': 'Excuse me, what is the name of that officer?' 'Lord Gort, sir!' someone exclaimed, not believing there was a soul who didn't know Lord Gort. 'Well,' said 'Puggy' (who had never heard of Lord Gort before), 'have you made arrangements to get Lord Gort off if the Germans come quick?' 'Oh, Lord Gort won't go,' was the reply. 'But it's your job to see that he *does* get off! It would be a disaster if your C-in-C was caught by the Germans. I shall leave a small vessel here for him to use.'

The Commodore discovered later that the other two officers in that group were Lord Alexander and Lord Montgomery: 'I found this out long after the war when Monty became President of the Portsmouth Branch of the Sea Cadets and I, as Hon. Commodore of the Sea Cadet Corps, had to go down and do the honours. There was quite a line of people to be presented and Monty said to most of them: "Now I don't think we've met before, have we?" When it got to my turn, I was all ready to reply: "No, I have not had the honour of meeting you before," when he said: "I remember you – at La Panne! You were all seaweedy and dirty and sandy." What an astounding thing! All the people he must have met during the war, and he not only remembered me but the clothes I was wearing. What a remarkable man!'

From GHQ at Dunkirk, the Commodore telephoned the Admiralty and asked them urgently to send more shallow-draft vessels, grass hawsers, snatch blocks and anchors to haul off tows from the beach. He requested the Embarkation General to run lorries into the sea, to act as piers when planked over.

Many thousands of soldiers were waiting to embark, and over several days and nights they were ferried or towed off in boats and transferred to whatever ship was available to take them.

One boatload of soldiers was particularly troublesome. It was a great big heavy boat – and, as it came near the ship, the soldiers all made for the side in order to get out quickly. Of course the boat canted over wildly.

'Sit down in the boat.' 'Puggy' said, 'Will you sit down in the boat!'

No response at all.

'You blasted sons of bitches, you lop-eared leopards, will you sit down or I'll shoot your heads off!'

That did it.

Now there were three parsons in that boatload, and when everyone was on board ship, 'Puggy' apologised to them for the language he'd used.

'My dear fellow,' one of them replied, 'I wish more of our officers would use such language – we might get something done!'

All this work had to be done under more or less constant shelling and air attack, which was sometimes very heavy, and, the Commodore was proud to report to the Admiralty later, with characteristic naval understatement, that 'the conduct and behaviour of all officers and men was in accordance with Service tradition'.

In that report five men were named who had behaved with special distinction: H.G. Hayes, W.A. Denny and E. Fenton, who volunteered to take a whaler ashore time and time again to bring off soldiers despite heavy shelling; A. Ferris, on duty alone in the twin-diesel engine room of HMS *Bounty*, who refused to leave his post though he was sick with the fumes; and Lieut. R.H. Irving, RNR, of the motor boat *Triton*, who set a tremendous example of devotion to duty, though he was obviously so tired he could hardly keep his eyes open. Later, Irving was awarded the DSC and Ferris the DSM.

But what of the Commodore's own conduct at Dunkirk? – at the age of 62. Little would be known about that if Lieut. Irving had not written a vivid report of his experiences on the beaches for his commanding officer at Sheerness. After describing a nightmarish crossing of the Channel by night in command of a wayward group of small boats, Irving went on: 'At 05:00 I arrived at the beach near La Panne, where there were hundreds upon hundreds of troops waiting to be taken off. I proceeded, stopped just out of the surf and commenced towing boatloads of soldiers to destroyers and trawlers.

'I continued until about 12:00 when, alongside a destroyer, a voice hailed me, saying "Well done, motor boat! Wait for me." An officer wearing a lambskin coat came on board. He was wet through; however, he said he wanted me to carry on as I was doing and that he would have one or two other jobs for me.

'Commodore Stephenson and I then carried on taking troops to ships . . . at 01:00 we rescued seven soldiers from drowning . . . continued rescue work until 02:00 approximately, when, being very close inshore, I went aground . . . but succeeded in floating her off about 04:00 and carried on as before. I was feeling the strain somewhat, but worse lay ahead. At about 08:00 there was a strong north wind and very heavy swell, the boat was rolling and pitching at every turn and going alongside ships was very difficult.

'Commodore Stephenson now asked me to take him out to a motor torpedo boat which had just come in, and he asked me to continue the good work alone . . . he said he could not allow me to rest, too much remained to be done. I might say that the Commodore said I was "a good fellow" and on other occasions said I was "a bloody fool".

'It is not my place to pass remarks about senior officers, but this officer, without cap, soaked through, without food, was a great example to me. He helped me to steer, to pass lines, to haul drowning soldiers on board and very often would say: "Come on the army! Where have I seen you before – you are so very good looking, I am sure I know you." At other times he cursed the lot as stupid.'

All this has the authentic Stephenson ring. Sir Gilbert's condition when he finally got home after Dunkirk was described by a reporter of the *Saffron Walden News*:

'The Admiral could not speak above a hoarse whisper due to shouting orders for three days and nights in the din of air, land and sea bombardments. He had been several times overboard during the operations and when he arrived in England he had only a monkey jacket and a pair of boots left out of all his personal equipment. At Victoria station he was mistaken for a refugee by a Press photographer. "It is fortunate for the man that I had lost my voice," said the Admiral with some heat!

"The vessels under my command," said the Admiral, "were the most varied collection that I should think had ever been brought together in one place, and seemed to include everything that floated on the face of the waters."

'As soon as he arrived the Admiral went ashore to find Lord Gort. However, the C-in-C would not leave until he had ensured the safety of as many of his men as possible. Eventually he was taken on board one of the launches under the Admiral's command and then transferred to another. This was just as well, for the first launch was almost immediately sunk.'

# 13

# *Western Isles* Under Way

## (1940)

With the Dunkirk interlude behind him, the Commodore was able to go up to the Clyde and check progress on the *Western Isles*. He had given the chairman of Barclay Curle a firm date for completion and told him that, short of invasion, pestilence or famine, he expected the date to be adhered to.

The day before completion date the ship was coaled and had steam up; 'Puggy' told the First Lieutenant they would sail at 9 a.m. next day.

'But we've got a lot of dockyard mateys on board,' he said. 'That's their affair. We're going to sail at 9 a.m. as I've told them.'

So they sailed with forty dockyard people on board – 'Puggy' wanted to teach the chairman a lesson. (He sent the men ashore further down the river!)

Then he anchored at Tail of the Bank. He noticed the pier was calling up. Neither of his signalmen could make head or tail of the message – in fact, it was just asking the ship to send a boat for mail.

So 'Puggy' ordered the starboard sea-boat to be lowered. Eventually the crew got away – but it was soon obvious that none of the men had ever used an oar before. They couldn't pull and, before they'd gone very far, the cry rang out: 'The boat's sinking!' So the Commodore told them to get it back on board. But the First Lieutenant had never used davits before, so *he* had to go and hoist it!

Then 'Puggy' said, not very hopefully, 'Lower the motor boat!' They got it into the water, but all the engine could do was grunt a few times. There was no petrol in it!

Next day, the engineer came and asked if they'd far to go that day because he was short of coal.

'But', said the Commodore, 'you reported to me that we were full up before we left!'

'Puggy' found it rather embarrassing to ask for more coal when he was supposed to have coaled the day before he sailed. But the engineer, he gradually concluded, was quite mad.

Soon afterwards the engineer came up again with a bit of metal in his hand.

'I'm afraid we can't sail till this is mended, sir,' he said, 'the starboard engine's out of action. It'll take about four days to repair.'

Now the Commodore was very anxious to get up to Tobermory, because there were already three ships waiting to begin training. Fortunately there was a helpful man from Barclay Curle on board who said he could get the engine fixed next day.

So the Commodore reported the damage, asked for tugs and got alongside the coaling wharf (not without a bit of damage to the starboard propeller!). At least it gave him an excuse to top up with coal without looking too stupid: 'There weren't enough dockyard people to coal us on their own so we all joined in. Some people were afraid we might cause a strike, but not a bit of it!'

Then the engineer came up and said the men were refusing to go to sea again – he said they were tired! By this time 'Puggy' was quite convinced he was mad. The Barclay Curle man went below to see what was wrong. He didn't really find out *why* they were tired, but advised the Commodore to delay for an hour to give them a rest.

The navigator did not exactly inspire confidence either. The Commodore called him in to discuss the passage to Tobermory and asked him what the tide was doing.

'Oh,' he replied, 'I don't take any notice of tides!'

Luckily the First Lieutenant knew when the ship was heading for the rocks.

Somehow they made it to Tobermory, but the setbacks were not over. They had ordered a water-boat to meet them. She was an old wooden drifter which had been with 'Puggy' in the Mediterranean during the First World War and unfortunately, on the way round to the Isle of Mull, she chose to collide with something bigger than herself and sank. So, at first, the *Western Isles* had to get under way and go alongside Macbrayne's pier at Tobermory each time she needed water.

There was great difficulty in getting proper water supplies fixed up. The town was short of water even for its own purposes. After studying the situation the Commodore came to the conclusion that the pipes from the reservoir ought to be enlarged, and proposed this should be done at the town's expense. They could then charge the Navy for its supplies: 'But they wanted to charge us such an exorbitant amount that I would have nothing to do with it and sought another way out.'

At one point in the bay there were some low cliffs with a trickle of water coming over them – produced by drainage from the hills around: 'I had this water very well tested and it was perfectly sound, though brownish in colour. We then built a dam to contain the water above the cliffs and made use of some air-trunking to convey it down to a small jetty we constructed for the ships to come alongside. This supply was adequate for our use the whole time we were at Tobermory. Of course, some of the men grumbled at first that they were going to be poisoned – but it was perfectly pure. And we'd paid nothing for it!

'The way we started that ship! There was everything against us-but somehow we made good.'

Once the Commodore and his small team had arrived at Tobermory they had to get to work without delay, and there was very little respite for the next five years.

How did they set about the task? It was defined as 'giving a refresher course of a week or ten days in new weapons and tactics to vessels of destroyer class and under, after long periods at sea'. With this object in view, only a very few staff people were appointed; no-one had any idea it was going to be such a big job.

Ships came to Tobermory in which the vast majority of the crew had never been to sea before – officers as well as men. They had to be taught the ABC of everything. Sometimes there were as few as three people with any sea-going experience at all. Most of them had never worked in a team before and a great problem was to get them to obey an order from the right person without hesitation: 'I don't think the Admiralty realised how raw many of the ships' companies were. They thought all we had to do was to "work-up" ships already commissioned, with crews who were fully disciplined and accustomed to working together. But we found this was very rarely the case. So one simply had to invent new methods. And, to be fair, the Admiralty never interfered.

'I had a rule never to ask permission to do anything, never to ask for anything – take it and tell them to pay for it, do it and tell them I'd done it! It may sound stupid, but I knew the Admiralty – it was a big organisation; no one man would deal with a paper, it had to go the rounds, it would take *days*. I'm not blaming the Admiralty – it couldn't be helped, and I was grateful for their co-operation in letting us carry on with the job.'

One example of Admiralty co-operation was over the question of a guard ship.

There never were any air raids at Tobermory – if German bombers got to Western Scotland they had Glasgow to go for. But it was an open anchorage and, when the Commodore first went there, he thought he ought to have a guard ship out at night in case of an attack by sea: 'that's what I ought to have done; but I thought if a ship was out on guard duty all night, she wouldn't be much use for training the next day; and we were up there not to guard the harbour but to teach – therefore we would have no guard ship!'

Then, one day, a vessel arrived to act as guard ship, rather to the Commodore's surprise and annoyance. He thought it a damned nuisance, so he asked the Admiralty 'on which days would it be safe to keep her in for boiler cleaning, and on which days should she go out?' She was withdrawn, much to 'Puggy's' relief. He didn't want another ship taking up room in the harbour!

One evening, quite soon after he began at Tobermory, the Commodore sat down and thought out his scheme of training: 'I decided we must have priorities – and my number one priority was *Spirit*: this was the first essential – determination to win. Next came *Discipline*: it's no good being the finest men in the world if you are not going to obey orders. Third – *Administration*: making sure the work of the ship was evenly divided; that meals were in the right place at the right time; that the whole organisation of the ship was both stable and elastic. Then, lastly – and this may surprise you – lastly, *Technique* – how to use the equipment. That would have been quite useless unless the spirit was right in the first place.'

'Puggy' got the discipline by close-order drill – getting people used to obeying new orders every five seconds or so. At first the gunnery instructors from *Western Isles* took each class, then the officers of visiting ships had a go. This taught them to give orders in such a way that they were obeyed. This was found very effective with all the many different types of ship that came into Tobermory Bay – escort and anti-submarine ships mostly, but quite a few other kinds, too. They all needed a similar pattern of training. It worked, in general – but not always without difficulty.

It surprised 'Puggy' to learn one day that a rescue tug was being sent to Tobermory: 'this was so unusual that I suspected it was either a mistake or a leg-pull. Anyway, I sent a young officer to Greenock, where she was fitting out, to find out all about it. When he got back, he told me she was coming to us, that she was entirely manned by ex-dockyard people, including her five officers, and that they were all absolutely determined that not even God Almighty was going to make them do close-order drill.

'Of course I had to get the better of them somehow. When she came in, I gave her a billet and told my staff that not one of them was to go near her without my express orders.

'I went on board as soon as she was moored, welcomed them to Tobermory, gave them the local orders about leave and so on, told them where the canteen was and where they could play football, and then left them.

'The next morning I went on board again about 11 o'clock to see the Captain. I said I hoped they'd had a pleasant first day and enjoyed a good run ashore. And then I asked him to show me round his fine ship, as I'd heard so much about it.

'We got to a place he called the Magazine – which looked as

though someone had just emptied a dust-cart into it, everything was just anyhow.

'"Oh, no", I said, "it's too bad of the dockyard to leave things in a state like this. Now I've got some chaps who are very good at tidying up and putting things in order. Would you like me to send one of them over to give you a hand?"

'"Well, thank you," said the Captain, "if it won't be putting you to too much trouble."

'"Oh, no," I said, "no bother at all. And by the way – I see you've got a gun here. Now I don't suppose it would interest you, but sometimes we secure a rifle to the barrel of guns like these – then a little boat is sent along towing a small target and your men could have a few pot shots at it to test their aim. Would you like this, do you think?

"Oh, yes, I think so – it's not a bad idea."

"All right, I'll send the rifle on board."

'The next day I didn't go near them till the afternoon; and when I went on board, I thought everyone was looking very bored and tired. This time I asked the Captain if I could talk to the officers and the ship's company.

'I told them we were very glad to see them and their fine ship with all its splendid gear. "But we've all been wondering," I went on, "how you manage to do all the things this ship is designed to do. In the dockyard each one of you has your job to do and you're left alone to do it. But in the work you've taken on now, the unexpected is always happening. One day a ship's torpedoed and you've got to get alongside, help strengthen her bulkheads, repair damage, take off the wounded; another day a ship's on fire and you've got to put it out; or perhaps a ship is sinking and you have to get on board and rescue as many as you can; or you have to pick up survivors from the water; or tow a ship. None of these things is very easy. Now what's going to happen on board when one of these things occurs, and when you're not yet trained to work together? As far as I can see, you're all going to collide with each other as you rush around in different directions."

'"Well," said one of the men, "what do you recommend?"

'"The great thing, as I see it," I continued, "is to have *one man* giving the orders, but, of course, that's difficult for you. You've always been accustomed to going your own way, and I hope you'll always be independent-minded. We don't want to change the Briton; but just now independence has to be forgotten for a bit and people have to learn to obey instantly. But, of course, it's not easy to learn new habits like this. Habit makes everything easy – for instance, on Friday nights your legs take you automatically into the 'Pig and Whistle' without your telling them. That's pure habit."

'"Well, how do we get this habit you're talking about?"

'"Well, sometimes we take about eighteen people ashore and we form them up and say 'Quick March', 'Halt', 'Right Turn', 'Left Turn', 'On Caps', 'Off Caps', 'About Turn', 'Sit Down', 'Stand Up', 'Stand on your heads'. It doesn't matter what the order is – it's just practice at obeying orders. That's what we do in the Navy."

'There was rather a pause and then someone said: "Can we do it, sir?"

'"Why not? Get the boats alongside, we'll go ashore and have a go at it!"

'You know, when they left us a fortnight later, they cheered ship. They went out to South Africa, and I heard later they swore by the Royal Navy!'

Fortunately there was less difficulty with most of the ships that came to Tobermory. But 'Puggy' soon became aware that he was regarded by many of his visitors as something of a terror – 'quite wrongly, of course, for I am, as you know, a nervous and timid creature and the mildest man you could hope to meet!' "The Terror of Tobermory" was a *complete* fiction – and the legends he inspired were for the most part quite unfounded in fact!'

However, 'Puggy' did nothing whatever to discourage them. It would have been no use if people had come to Tobermory feeling that there was an amiable old chap in charge who just had to be buttered up a bit and all would be well: 'Remember – these fellows had absolutely no knowledge of the sea and, in a fortnight, had to learn what it had taken me between five and ten years to learn!' Admiral Sir Caspar John, when he was First Sea Lord much later, told Sir Gilbert Stephenson that the Navy's 'working-up' procedures were based entirely on his methods at Tobermory.

To train raw recruits (as many of the visiting crews were) up

# 14

# Legendary Commodore

O f the stories people remember about the Terror of
Tobermory, it is difficult to find many that have not been
at least a little embroidered in the telling, but, whether true,
fairly true, or downright fiction, they do give a picture of the
man as others saw him – or thought they saw him; taken together
(and many more stories are told than we have room to print in
this book) they suggest how and why Sir Gilbert became a legend
in his lifetime.

To begin with first impressions: what was it like to arrive for a
fortnight's work-up at Tobermory, with the legend probably
imprinted on your mind before you began?

'When a new ship noses into the harbour,' wrote a Liverpool
reporter, 'signals flash from the bridge of HMS *Western Isles*. The
Commodore's barge plumes across the water and no sooner is
the ship anchored than her Commanding Officer is on his way
to the flagship. Once the Commodore asked a Commanding
Officer what he had heard about the base. "Well, frankly sir,"
replied the CO, "I've heard you're tortured until you're
efficient. Then, if you've any strength left, you're allowed to
go!" '

This was a civilised encounter compared with the welcome
accorded to three motor launches which arrived at Tobermory
in March 1943 – if one is to believe the recollections of Lt.-Cdr.
C.A. Head, DSC, RNVR:

'Stories of the legendary Commodore reached us at every port of call, gaining in horror as the distance from Tobermory decreased. We had left Fort William keyed up and apprehensive, and feeling anything but "ready in all respects to meet the enemy".

'On arrival, 559 was signalled to make fast to a buoy off the pier, with the other two boats on either side of us. Hardly had this manoeuvre been accomplished when the Commodore's barge was seen speeding across the harbour in our direction. This was it! Welcoming parties, complete with bos'n's calls, were mustered on each of the outside craft ready to simulate a Naval welcome. Nervous glances were cast around the upper decks, the crews looked even more amateur and inexperienced than usual, fellow officers suddenly failed to inspire confidence.

'The barge, very close now, seemed to disappear. Its engine cut. Where was it? Suddenly, over the flare of the bows of 559, in the centre, appeared a bewhiskered face, visible from the bridge but invisible to an Ordinary Seaman leaning on a broom, his eyes fixed on the bos'n's parties waiting, and looking puzzled, on the outer craft. The Commodore now had his knees on the gunwale. He hauled himself up, straddled the guardrails, slapped the Ordinary Seaman on the back, picked up the broom he had dropped, and, holding it out in front of him like a boarding pike, charged at the bridge.

'As I froze into a tremulous salute, with one eye on the broom, the Commodore barked: "I want the mincer – and your ship is on fire in the engine-room!"

'After a panic of hesitation in which I nearly sent to the engine-room for the mincer and instituted fire-fighting operations in the galley, I started to give what I hoped was an appropriate series of orders. Hoses were rigged and began discharging water, foam extinguishers were produced, a stoker stood by in the wheel-house ready to pull the lever which would fill the engine-room with lethal methyl bromide gas. Still the Commodore stamped the deck, shouting that the fire was spreading.

'At this point I presented him with the mincer. It was still wrapped in the tissue paper in which it had been issued three weeks before from Naval stores. The Commodore sized this up straight away and, realising it could not possibly be unclean, waved it aside disdainfully. I continued to exercise my command clutching the mincer like some curious badge of office.

'Meanwhile officers and crews of the craft alongside watched delightedly the entrancing show which was being directed by the Commodore on 559, apparently for their especial benefit. Witty but muffled encouragement passed over the guardrails; some suggestions, too, not all of them serious, or meant for practical application. Meantime, the Commodore continued to direct the fire into new corners of the fabric.

'Now, if there is one thing more difficult to handle on board ship than a genuine fire, it is an imaginary one, especially if it proceeds from an imagination of endless resource and malevolence. Should one actually direct the water jets down the engine-room hatch? Should one start the foam? Should one release the poisonous methyl bromide? I was getting desperate. If I was to be relieved of my command and returned to base, I might as well be hanged for a sheep as a lamb. I was about to give the irrevocable commands. However, the Commodore, though seemingly lost in a paroxysm of rage, had evidently been keeping a very cool and experienced eye on me, expecting this development. He quickly turned on his peroration: "My God! This ship is going to be a total loss. She'll go up any minute now, and" – he looked balefully at the officers on the next ship – "by Heavens, the whole damned lot will go up with her!"

'I realised now that the whole point of the inextinguishability of our private fire had not been to see what *we* should do. The Commodore was not interested in us at all; his target was the double audience of grinning onlookers to starboard and to port. The neighbouring COs felt the fiery gaze swing on them; I felt it switch off me. I now had time to put the mincer down. The other boats cast off in a hurry and began to circle the bay aimlessly, left to their own devices, not knowing what they were expected to do now. The Commodore was apparently not interested any more.

'Having exploited us for this ulterior motive, the Commodore was now chatting to me affably; the fire-fighting gear was being stowed away and all seemed to be over. Unfortunately the First Lieutenant chose to relax within view. "Away ship's dinghy . . . Under sail!" said the Commodore benevolently.

'No command could have been less welcome. We had not thought of this one. The fact that the Navy had provided our tiny dinghy with sails and rigging had seemed to us merely an amusing anachronism. They were stowed deep down. Did anybody know how to rig the wretched boat and sail it? I had no idea. And still the Commodore engaged me in charming and irrelevant conversation. It was the final test of nerve. Luckily one of the seamen had sailed as a hobby as a boy. He stood at No. 1's elbow, whispering counsel and, miraculously, in a fairly short space of time the dinghy left the ship for some unknown destination.

'The Commodore, relenting, concluded that this was enough to start with, and, shouting: "To your tents, O Israel!" vanished with amazing alacrity over the bows into the void, whence he emerged a few seconds later sitting composedly in his barge. We watched it go round to the far side of *Western Isles*. Only then did anyone feel safe.'

Sir Gilbert, though much entertained by this narrative, firmly denies that he ever boarded or disembarked from a vessel in any such fantastic fashion, but he admits that surprise was often a feature of his technique: 'as if I would ever climb up the bows to get on board a ship! Sometimes I did go out to meet them before they'd actually secured, however, and shout out: "Throw over a Jacob's ladder" – and I'd go up that. It was to impress them with a feeling of urgency. This was one of the things I wanted to convey.'

The Captain of one ship was at breakfast one morning when the Commodore's barge arrived to bring him to *Western Isles*. 'Oh, I'll just finish my breakfast – won't take a minute or two,' said the Captain. 'Well, sir, if I were you I wouldn't keep the Commodore waiting . . .' advised the experienced Officer of the Day. When the Captain returned from his visit he said to the First Lieutenant: 'My God, never keep that bloody fellow's barge waiting one second if you can help it – you don't know what he said to me. I've never had such appalling things said to me in all my life, and I hope I never will again.'

Another Captain, Commander R.H. Bristowe, was having a haircut in the Coxswain's office of HMS *Castleton*, an American Lease-Lend destroyer, at Tobermory, one morning in May 1942:

'Halfway through the operation, a young Ordinary Seaman, unused to the sea, came into the office and said: "Excuse me, sir, there's a gentleman to see you on the upper deck". This surprised me, in the wilds of Scotland, until he continued: "He's in uniform and doesn't look pleased, sir". The truth dawned; the Admiral had "warmed the bell". With half my head shaven I dashed on deck to find the Terror of Tobermory on the quarter-deck with my First Lieutenant and living up to my OS's description; he was *not* looking pleased.

'He attacked me at once, covered in gold braid and decorations. "I come on board a British man-o'-war, with the Empire in the throes of battle, to find the Commanding Officer having his hair cut and his First Lieutenant eating a bar of chocolate!" (Which he was still in fact doing.)'

Bristowe was by no means alone in being found thus unprepared. Once, the Commodore boarded a corvette, and seeing little sign of life, made his way to the wardroom, where he found several somnolent officers. The First Lieutenant jumped to his feet.

The Commodore enquired where the ship's company were employed, and was told that they were painting the ship's side. 'Right,' said the Commodore, 'let's go and see.' On the fo'c'sle there was no sign of a sailor to be seen, so the First Lieutenant said they must be painting over the stern. Still no sign of any men. 'I'll tell you where they are,' said Sir G., 'let's go to the messdeck.' Here were found numerous sleepy seamen. The Commodore turned to the First Lieutenant. 'How many store rooms have you in this ship?' No. 1 told him. 'Right, bring all your stores up on deck and, when you have done this, make me a signal. I will then order you to strike them down again!'

In the 'Terror's' view, the efficiency of a ship depended above all on the efficiency of her officers, so he reserved most of his ferocity for the incompetent among them. However, his impact on the lower deck was hardly less immediate – on Michael Low, for instance, an AB in the Flower Class corvette HMS *Snowdrop*.

'I remember him from his red seafaring face with hair growing from below his cheekbone down to the middle of each cheek. His eyes were steel blue, deep piercing and, to me, frightening.

'I encountered the famous "Monkey Brand" when I was

washing down the paintwork in a narrow passage outside the engine room, the day after receiving a black eye from a fall down the steps of the bridge. I was met by the famous man who growled at me: "If I see another person on this ship with a black eye, you are *both* on a charge!" I stammered: "Yes, sir" and he was on his way, but I will never forget his face.'

At least the lower deck were spared one of Tobermory's most alarming ordeals – the invariable invitation to Commanding Officers to dine in HMS *Western Isles* on their first evening at the base – [an occasion widely dreaded, with varying degrees of justification. When the RN Captain of the frigate HMS *Rupert* arrived in the Commodore's cabin, a signalman was at once commanded to appear. The Commodore then dictated this message: 'To *Rupert*. Lower your motor boat and proceed to *Western Isles* without using the engine.'

Fortunately a signalman still remained on the bridge of *Rupert* and took down the signal. This was handed to the First Lieutenant, a young RNR Officer, who immediately thought it was some sort of a joke. After some thought, he decided he had better take some action, and the boat was eventually lowered. Propulsion was the next thing to be thought about, and finally he hit on the idea of using Carley float paddles. This was duly done and the motor boat proceeded shorewards in the fashion of an Indian war canoe! All this time *Rupert* had been under observation by the Admiral who, consulting his watch, said: 'Fifteen minutes, Captain. Far too long for that operation!'

Alas, poor *Rupert*! What is worse, Sir Gilbert Stephenson's version of this story is even less flattering to the ship and her Commanding Officer:

'I made a signal to a ship instructing the Captain to come and see me; the reply came back: "Regret boat's engine broken down; hope to have it repaired in four hours." I sent a second signal: "Captain repair on board," to which the ship replied: "Hope to have the boat ready within two hours." I then made a third signal: "If the Captain is not on board within twenty minutes, I shall send a file of Marines to arrest him." The motor boat then arrived with the Captain on board, paddled by sailors and stokers!'

The Commodore always reserved his sternest tests for Royal

Navy – rather than Reserve-Officers, for the professionals, he felt, should be resourceful enough for the most outlandish emergency. So a variation on the 'boat' theme was laid on for the benefit of a submarine CO – another Royal Navy man. (Submarines were sent to Tobermory to play the part of U-boats for training purposes, and there are submariners of that time who allege that the Commodore knew nothing whatever about them – going so far as to assert that 'on one occasion he gave a submarine crew time off for boiler cleaning!')

In accordance with custom, the Captain of this particular sub was invited to dine with the Commodore, and consternation broke out when a signal arrived stating: 'Shall send whaler to collect you. You will provide the crew.' Now submariners are not particularly renowned as skilled oarsmen: however, a scratch crew was quickly assembled and off went the Captain.

Unfortunately, when the whaler arrived at the *Western Isles* it became hopelessly trapped under the stem of the Commodore's ship. The Commodore, who was awaiting the arrival of his guest on the quarterdeck, eventually could stand it no longer and, craning over the stern, shouted to the embarrassed submarine skipper: 'Lt. —, I asked you to come to dinner – not bloody breakfast!'

The wind, however, was often tempered, even unto RN Commanding Officers like Commander Ralph Jenkins, OBE, DSC, who arrived at Tobermory exhausted and was appalled to receive the ritual invitation:

'I was in no mood to be sociable and had every intention of getting my head down without delay. However, the Duty Staff Officer in HMS *Western Isles* was horrified at the idea of a visiting Commanding Officer "Under Training" declining the Commodore's invitation, and made it quite clear that this was in fact a sort of Royal Command.

'So I shifted into mess kit and joined, albeit rather late, Commodore Sir Gilbert Stephenson for a most pleasant and, as it happened, restful dinner which included English artichokes grown by the Commodore in his Scottish garden. The Commodore was no fool. He was well aware that an early night was what I needed, and he sent me back to my ship in his barge at not too late an hour.'

Lieut.-Cdr. E.F. Aikman, RNR, in command of his first frigate, also had a pleasant surprise:

'On arrival at Tobermory I had to make fast to a buoy which was about twice my ship's length from the rocks, on a dead lee shore. When I went on board *Western Isles* to report to the Commodore, he complimented me on having made fast in quick time. This surprised me, as I thought we had been slow, but no doubt he was taking the state of my ship's company – 70 per cent of them new to the sea – into account. I remarked on the difficulty of holding the ship's head to wind under such conditions until the picking-up wire was secured, whereupon he gave me one of the best pieces of advice I ever received on ship handling. "Why don't you try coming downwind to the buoy? I know it is against all your instinct as a seaman to approach your buoy before the wind, but try it." As I subsequently found, he was absolutely right.'

This is just one example of the profound understanding that lay below the Commodore's daunting exterior. Fundamentally, he knew how immense were the problems that faced these raw young men, how outrageous were the demands made on their inexperience by the war.

Just occasionally, the essential humility of the man showed through – as when he remarked quietly to one young Captain: 'You know, I tell you fellows to do this and that, but I sometimes wonder how I would do it myself!'

# 15

# Admiral Brand

L et us now pursue the tale of Tobermory beyond the ordeals of arrival and first encounter with the 'Terror'. What did the Commodore demand of his victims in the course of their training?

In general, the first few days were relatively calm, though full of hard work for all. As the sun rose – or before – classes left their ships for instruction on board HMS *Western Isles* in subjects like aircraft recognition, anti-submarine drills, spotting and reporting and signals; while in the visiting destroyers, frigates, corvettes, sloops *et alia*, those left behind would be sweating over gun and depth charge drills, Asdic, communications, and many other exercises, under the supervision of *Western Isles* staff – and the ship's officers had to find time to draft elaborate fighting instructions and standing and departmental orders as well.

It was just when all this was coming along quite nicely and the visitors were getting lulled into a sense of security that the Commodore would stage one of his highly dramatic 'emergencies'. (Though it could happen on the very first day – witness the MLs in the previous chapter.)

The Commodore or one of his team would come on board a ship and inform the Captain that his ship was sinking – and on fire; that he was to engage enemy aircraft with his four-inch guns and prepare to fire depth-charges at the same moment – despite the fact that his electrical power supplies had been shot away and all his officers were dead. 'Improvise, my boy, improvise,' the Commodore would shout cheerfully, while lighting a Thunderflash from his pocket and flinging it down behind anyone who was not moving quickly enough.

One Commanding Officer was happily thinking he had escaped this treatment when he was summoned on board *Western Isles*, where a children's party was in progress. He was introduced to each child individually, and just as he was wondering what games he should think up for them, the Commodore gently led him to the ship's rails.

'See those hills over there?' he asked. 'Well, the town's water supply lies behind them and paratroops have just dropped to destroy it. Deal with the situation, will you, and maintain contact with your landing party . . . All right, what are you waiting for then?!'

Sometimes the Commodore would board a ship and order the Captain to get under way. Just as he was slipping his moorings would come the massive 'enemy air attack', followed by an order to 'send a landing party ashore'. Then, in the middle of the crowded harbour, the Captain would be told that his steering had been shot away and his bridge completely destroyed – he would have to go down and con his ship from aft. Whatever was happening, the Commodore's eyes and ears were very sharp – he once ordered the Yeoman of Signals on board a frigate to send a certain message by lamp, and in the midst of the considerable chaos raging around him, noticed that the Yeoman had misspelt a certain word.

A man as quick as the Commodore was rarely outwitted; when he was, he usually reacted with great good humour. Once, when he was putting a trawler through its paces, he said to a Leading Seaman, a member of a gun crew: 'You're dead, lie down.' The rating obeyed but, after a few minutes, the Commodore noticed he was on his feet at his usual station and, in his usual fierce manner, exclaimed: 'I said you're dead!' The rating's reply was: 'Dead or alive my place is at this gun'. The Commodore's reaction? 'Full marks.' And a young RNVR officer was commended for his response when the Commodore came into the wardroom one day and said to him: 'There's a fire on the quarterdeck!' To which the officer replied: 'Do you mean to say, sir, you've done nothing about it but come down here and tell me!'

What happened if a ship failed to measure up to these schizophrenic situations? The Commodore was not by any means beyond removing a Captain or First Lieutenant he thought not equal to the task – or the ship could be threatened with an extra week at the base – or the reaction might be more immediate, as was the case with one poor corvette.

The first day of their sea exercises included towing and being taken in tow by another escort vessel. The anchor was detached, three shackles of cable paid out to form a towing spring, and the manoeuvre successfully carried out: that is until the time came to recover the cable. By then the wind was increasing and the notoriously lively antics of a corvette did nothing to help what little seamanship the ship's company had. At any rate, in some way perhaps best forgotten, there was a rattle and roar – and three shackles of valuable cable lay on the ocean floor.

The Captain, after forcefully expressing his opinion of those concerned, ordered the spot to be marked with a buoy – but, because he had an appointment to keep with a submarine for the afternoon's exercises, decided to abandon any attempt to recover the cable, which was a pretty hopeless task anyway. The loss was reported to *Western Isles* by signal. The weather continued to deteriorate, and by the time the corvette returned to harbour in the late afternoon, a gale was blowing; the approach to her mooring buoy was a long and arduous task, and it was only after much backing and filling that she at last secured, with everyone more than ready to relax after a hard day's work.

But they had reckoned without the Commodore. No sooner had the pin of the mooring shackle been driven home than his barge was cutting a swathe towards the corvette; he stormed up the gangway bristling with indignation.

'Lost your cable?' he barked.

The Captain answered rather shamefacedly that he had.

'Then what the devil are you doing in port? If you'd lost your grandmother over the side you'd still be out there looking for her, wouldn't you? Eh? Wouldn't you?'

The Captain agreed that he would. It seemed safer to do so.

'Then out you go and look for your cable – and don't let me see you here in the morning.'

And out into the night and the gale went a very tired and unhappy corvette to drag a grapnel up and down for weary hours before the Commodore relented and sent her to other tasks. But her ship's company had learned a lesson – it was no good giving up easily – not when you were dealing with the Terror of Tobermory.

Visitors to the base were most unlikely to get to windward of the Terror; however, Captain C.F. Vine, DSC, RD, RNR, claims to have done so, because his brother, who'd been to Tobermory already, had forewarned him.

Vine therefore knew that the Commodore wanted all hands ashore to hear his inaugural address, and that most ships were in trouble for leaving their cooks and stewards on board. Needless to say, he had all his cooks and stewards in the audience. He knew the Commodore hated oilskins hanging in the messdecks, so removed them in advance. And he was prepared for the exercise of shoring up a bulkhead. This, he was told, had to be done without cutting the timber, with the result, generally speaking, that when the Commodore gave it a kick the whole lot would collapse, provoking the remark: 'Do you call this shoring up a bulkhead?' Having been warned, Vine had his timber cut to size before the ship arrived, so when the Commodore gave his expected kick, the bulkhead, in return, gave his toe a nasty shock.

Yes, just occasionally the 'Terror's' technique misfired. As when Commander Gordon Steele was bidden to accompany the Commodore on one of his visitations:

'About 8 a.m. he gave orders to his coxswain to proceed alongside "that ship" – pointing to a frigate second in the line of some six at anchor. We arrived alongside, and the Commodore sprang up the gangway, with the agility of a man twenty years under his age, to find no one to greet him – they were probably at breakfast. He stormed; a couple of frightened hands on deck rushed below, presumably to call the Captain and the Officer of the Day, for they appeared, breathless with alarm. They were soundly rated for their slackness in not being on deck to receive him, and then the Commodore got down to real business with the Captain – an RNVR Lieutenant-Commander.

'It appeared that numerous items in the course of Tobermory training had gone wrong with that particular ship; there was a long list of failings, from omitting to receive signals to taking up the wrong station. The Commodore continued, breathless. Once or twice the Captain attempted to interrupt, only to be shut up. I stood back a few paces, feeling extremely sorry for the man. However, the Commodore eventually came to the end, exhausted at such expenditure of "steam". He turned to go down the gangway and only then could the Captain of the frigate get out the words: "I am afraid, sir, you have come to the wrong ship!" And so he had.

'It was still too early for the Commodore to laugh at himself, which he so frequently did, and which so accounted for his popularity. He just said: "Return alongside, coxswain" – and called it a day.'

Another good story of the Commodore getting his comeuppance tells of the time when he went on the bridge of a ship and pressed the alarm bell for a long time. The expected eruption of activity was lacking, and eventually the Chief Engineer came up through a forward hatch and said: 'If you don't stop pressing that bloody bell, you'll run the battery right down!'

In his desire to stimulate invention and initiative, the Commodore developed a whole repertoire of ingenious tricks.

He would send a rating into the crow's nest and then tell some poor Sub-Lieutenant or Snotty that the man was up there with two broken legs, and must be got safely down. Or people normally well removed from the sea, like cooks and stewards, would be told to build a raft – this was a favourite test. On one occasion the Commodore signalled a motor launch to construct a raft and sail 'round the Western Isles'.

After some thought, the crew decided he probably meant the headquarters ship and not the Hebrides, but in any case the raft sank gently almost as soon as it was launched!

New and unexpected challenges were invented and flung down day in and day out for five years. Nor did battle cease at nightfall. The Commodore devised an original plan for keeping

Quartermasters and lookouts on the *qui vive* during the dark hours.

A boat with muffled oars, or a quiet motor boat, would be sent out from *Western Isles* with instructions to approach this or that ship – and if they were not duly hailed, to board her and try to remove some vital piece of equipment. Next morning, the night's haul would be displayed at the yardarm of *Western Isles* and Commanding Officers would be summoned to collect the missing articles.

Sometimes the approach was slightly more subtle.

One night, accompanied by his coxswain and wearing an AB's cap, the Commodore personally boarded a trawler, and the pair of them manhandled a small safe containing the confidential papers from the skipper's cabin into their rowing boat and took it back to the base-ship. In the morning, the Commodore ordered COs of ships in harbour to repair on board in order to have their confidential books checked.

Imagine the scene – the unfortunate skipper, whose birthday it had been the previous day, no doubt suffering from the after-effects of a party, coming along to report that he could not find his safe – let alone his 'CBs'.

Commodore: 'Do you always keep a good look-out whilst at anchor?'

Skipper: 'Yes, sir.'

Commodore: 'B— lies. How do you account for the fact that I am in possession of your safe?'

It was perhaps hardly surprising that a number of ships took revenge in one way or another for these nocturnal skylarks.

One morning a motor boat from *Western Isles* was found secured to the stern of a French corvette – Frenchmen can use muffled oars too. A Canadian destroyer, though the Commodore strongly denies it, claimed to have removed his barge and gone to sea with it. And in a particularly bold coup, some officers from a Canadian frigate actually 'captured' the *Western Isles* one night, as Commander Steele remembers:

'There seemed nothing to disturb the peace and silence, as a motor boat, proceeding unlighted and at a noiseless dead slow, approached the flagship's port bow; it stopped, and up the cables clambered some five or six dark, armed figures. In no time they had crept along the shadows of the upper deck and then aft to the Admiral's sleeping cabin, where – zero hour – some half-dozen warriors shone dazzling torches on the Admiral's sleeping countenance, pointed revolvers and demanded his immediate surrender!

'They could not tell how he would react; he lay still for a few seconds, with his brain working feverishly, no doubt. But he got out of bed quietly, still covered by the torchlight and the revolvers, and went to a small writing table on which lay handy for emergency the orders for the night – which had instructed certain ships to invent an imaginary objective for a nocturnal raiding party – and to send such a party off to carry it out. The Commodore perused the orders for a moment in silence – without doubt to ascertain if there was a loophole which could save him from the situation; apparently satisfying himself there was not, he beamed at his captors, to their immense relief, and, going to a small cupboard, exclaimed: "Have a drink".'

This game of night raiding eventually came to an end, probably because it became altogether too much of a good thing – or perhaps, as J. Ivester Lloyd related in his book *Wavy Navy*, the ships had got wise to it:

'There was the tough AB in a frigate who, when asked whether anything unusual had happened in his spell on deck in the middle watch, replied: "No, sir, nothing much. Only some bastard stuck 'is 'ead up over the after rail." When asked what action he took, the AB calmly said: "Flashed me torch in 'is face, then 'it 'im over the 'ead wiv it. Didn't see 'im no more." '

There was also a corvette which kept a firehose rigged in readiness and, when someone tried to scramble aboard unannounced, let him have it full blast. When the Officer of the Deck looked over the side he saw the Commodore himself, drenched to the skin and dancing up and down in his launch, laughing and saying 'Well done – oh well done!'

Maybe the Commodore's strange tactics had, after all, got his message across!

In all aspects of the training, his sense of humour was often in evidence. Clear communication being a prime objective, he made a point of exposing obscure or verbose signals – as when one ship sent in a particularly rambling request for certain supplies. In response, the Commodore dictated the reply – 'These will be sent yesterday. Top of the box marked bottom to avoid confusion'.

But an essential kindness never lay far below the surface. 'Monkey' would go to great lengths, for instance, to organise special transport for a man who needed compassionate leave – and he showed the closest concern for the welfare of his staff. In fact, whatever degree of terror he may have inspired at Tobermory, the Commodore is remembered with an affection which far outbalances it. When it was time for the trainees to leave the base, many did so with a tinge of regret mixed in with their undisguised relief. Lieut.-Cdr. Alexander Jarvie, DSC, RNVR, of the Fleet Minesweeper HMS *Wave*:

'The passing out inspection was chaotic and I remember feeling quite limp when we received our sailing signal and knew we had "made it". But I was very moved when the Commodore, in saying goodbye to me, said: "First Lieutenant, I envy you taking this fine young ship's company to sea". You knew instinctively that he was genuine – that he'd have liked nothing better than to go to sea himself. That not being possible, he did the next best thing in turning out others fit to do so.'

And Alexander Kirk, a former coder, recalls a farewell parade on a very wet day: 'We had to turn out in our "number ones", no coats; our feelings were not exactly happy and were expressed in usual Navy language, but, when the Commodore addressed us, standing in the pouring rain without any protection, I, at least, admired him very much and realised that the Terror was one of the reasons for the greatness of the Royal Navy.'

A truth expressed in one form or another by everyone who encountered the Commodore. But for all the impact he made, it was possible for at least one officer (still in a condition of delayed Tobermory shock?) to be delightfully vague about the whole thing:

'I remember we were put through our paces by an Admiral Brand, who was affectionately referred to as "Monkey Brand". I do not recall Commodore Stephenson, but presume he was succeeded in the post by Admiral Brand!'

# 16

# How it was Done

Unconventional though the Commodore's approach was, the theory behind it was that of all 'work-ups' for fighting ships – they have to be put, in imagination, through every conceivable situation which may face them on active service. In the course of this process, a ship and her company evolve methods of coping with whatever may come their way, based on individual specialist functions, and on the growing smoothness of the team as a whole. As we have said, all this normally takes considerable time – the peacetime Royal Navy allows several months for it.

But in the Second World War, there was often very little gap between a ship being launched and being involved in active operations at sea. The thousands of little ships needed to guard our convoys were rushed through the shipyards and commissioned at top speed with officers and men who only a few weeks before had been at an office desk, a factory lathe, at the university, or at school. A crash course in Naval ways at a basic training establishment such as HMS *Ganges* at Ipswich, mainly under elderly petty officers whose spiritual home was a battleship of the First World War, had to suffice before we went off to some little minesweeper or frigate as new as ourselves to the sea.

It was common for these ships to start their lives with no more than ten per cent of experienced officers and men. The regulars were hard put to it to be patient with our ignorance – and we, essentially civilians dressed as sailors, were not always tolerant of Royal Navy niceties. It was in this uneasy condition that most small ships were sent off for their 'work-up'.

There were several bases for this purpose around the British Isles, regarded by the RNVR, as J. Ivester Lloyd puts it, 'in rather the same light as that in which early Christians looked upon the Roman circus'. Some of them seemed to their victims to be no more than a safety valve for disgruntled retired officers to work off an excessive knowledge of King's Regulations and Admiralty Instructions; but Tobermory was different. In their book on the RNVR, J. Lennox Kerr and Wilfrid Granville hit on the secret of the Commodore's success:

'He realised that he was handling civilians trying to become sailors and did not insist – as so many of the older Naval officers did – on those 'Naval practices' which, admirable though they were in peacetime and on board ships carrying crews numbering some hundreds, are not suitable to smaller ships. He tactfully ignored, if he ever noticed, that the crews of HM Trawlers did not enter harbour lined stiffly on the forecastle head, or in their correct 'Number Threes', but he came down like a fiery torch on dirty messdecks, untidy boat gear, or other signs of carelessness or laziness which might lose a ship or drown a crew . . .

'The aim was the making of an efficient, alert fighting ship, where that old sailor's gospel: "The ship first, shipmates next, yourself last" was the inspiration.'

It was to get across these simple principles in a very few days, that the Commodore devised his shock tactics. They were all of a piece with a career never marked by meek submissiveness, and they brilliantly achieved their aim.

The only trouble, perhaps, with so individual an approach was that the whole operation – indeed the whole of Tobermory – came to revolve around this one remarkable man – in his view to an excessive degree, for he had no real deputy, and his absences left a painful gap.

This is not to say, of course, that the Commodore did not have able assistance. The First Lieutenant he had chosen at the outset for HMS *Western Isles*, Lieut.-Cdr. Palmer, was admirable, though his responsibilities did not extend beyond the headquarters ship herself.

In 1941, the Admiralty replaced the first *Western Isles* with a larger vessel – an old Belgian packet – and sent the Commodore their ideas for her conversion. His first reaction was to tell the Admiralty they'd have to get someone else to man the ship, for he wouldn't do it: 'No one apparently had thought how the thing was going to work – how the meat was to be got down to the cold store, for instance. And they'd put the Petty Officers' mess next door to my cabin, with only a thin bulkhead between: I wouldn't have minded very much, but the Petty Officers would have seen themselves dead before accepting such a situation!'

The Admiralty responded to the Commodore's strictures by asking if he'd like to plan the ship himself. He agreed, but handed the work over to Palmer, who did a perfect job: 'There was even a place for the "snob" – the boot repairer!'

The Commodore's main concern, of course, was with the training programme; but for all that, he took a very direct interest in the domestic arrangements of his own ship. He was said to have a pet aversion towards supply officers – and when a new one, James Walker, arrived in 1945, his departing predecessor wished him luck, saying that he himself was already the 'umpteenth.

According to Walker, the Commodore 'concerned himself with everything and everybody and nothing was too small to escape his notice – he even questioned me thoroughly over a reported shortage of soap. On his rounds of inspection he noticed that rats were getting at some of the stores and I was told to get rid of these pests. I did my best with wire netting and poison, but rats are cunning creatures and still the odd one managed to get in. This did not satisfy him, and I remember his remark to this day: "Walker, are you going to beat the rats or allow the rats to beat you?"

'You could not pull the wool over the Commodore's eyes, but you could always count on his support if you had just cause. I had difficulties due to shortage of staff, and work was getting behind. In desperation I approached the Commodore. He sent for the Warrant Supply Officer from Oban to report upon the position to him. Fortunately, the Warrant Officer agreed with me, and, as a result, the Commodore sent the following signal to the Flag Officer in Charge, Greenock:

' "As a result of the visit of Mr G.E. Sells, Warrant Supply Officer RN from Oban, I am advised that the staff which are at present available for duty are quite insufficient to keep the victualling accounts.

' "Under these circumstances it is forced on me to maintain the essential work of demanding and issuing the victualling stores, and for the *accounting to lapse*(!) until the personnel employed are again back to complement."

'It was an unheard of thing to let all accounting lapse!'

It was by such means that the Commodore kept his superiors in order.

Another crucial engagement was fought when he was informed by the Admiralty that, as *Western Isles* was a depot ship in Tobermory Bay, they would have to cease the allowance of duty-free drink made to sea-going vessels: 'This gave me considerable anxiety because I knew the effect it would have. So, a few days later, I arranged for the *Western Isles* to leave her buoy and go out on an exercise.

'The day after that exercise I wrote to the Admiralty: "I am afraid you are under a complete misapprehension about *Western Isles*; only yesterday she was at sea conducting a large exercise in the open waters off the west coast of Scotland". What an upheaval! I was very anxious about taking her out to sea because she'd had all sorts of gear put on top of her, and she'd had no ballast added to counterbalance it, so I was afraid she might topple right over. She was completely unsafe; when we came back I had a lot of the gear taken right down below in case we went out again. But once was enough – we kept our duty-free liquor!'

By such resourceful means, the Commodore became monarch of all he surveyed in Tobermory Bay – his head-quarters ship, the visiting ships under training – and the little town of Tobermory itself.

Soon after he first arrived, he went ashore and called at the manse to talk with the minister, Mr Menzies, and his wife Emma. He asked, first, for a playing field for the men in his care, and knowing of Scottish susceptibilities regarding the Sabbath, got an assurance that they would be free to use it on Sundays. Then he turned to Mrs Menzies: 'I want you to call the women together and organise a canteen which will be open every day of the week'.

Mrs Menzies said she would call a meeting 'tomorrow'. The Commodore roared: 'I don't like the word "tomorrow". Call it today.'

The meeting was duly called. Sixty women attended and all put their names down as workers in the canteen, and, for the five war years, those women continued to work on their four-hour shifts with no quarrelling, no interference, and nothing but thanks and gratitude for their voluntary work from the whole Naval squadron.

The third thing the Commodore wanted, that day he called at the manse, was a Sunday service. 'Will you come to the ship or shall we come ashore?' he asked the minister.

'Which would you prefer?'

'Oh, the men would far rather come ashore. It's a change for them and they like to see the girls in the choir, so put the pretty ones in the front row.'

So they went ashore – every Sunday – and the service was held an hour earlier to suit them. The Commodore read the lessons. One Sunday he began: 'El Alamein has fallen! Praise the Lord!' And on another, when he announced the loss of the *Hood*, the silence that fell is still remembered.

The Commodore liked the men from the visiting ships to attend the services too. One Sunday, the officers from a ship which had just come in were conspicuous by their absence.

'Why are the officers of the —— not here?' he asked, 'Are they all bloody heathens?'

Mrs Menzies started a choir which met in the manse for practice once a week. The men loved it, as it meant sitting in a comfortable chair by a fire; it also meant late leave for the men and Mrs Menzies would sometimes ring the Commodore to ask for this privilege. She would begin an explanation: 'You see, we can't be finished in time for them to get the liberty boat'. He would stop her: 'If you say they are to get late leave that is enough for me. Don't waste time giving reasons.'

(Sometimes, when Mrs Menzies phoned, the Commodore would shout, 'What language are you speaking?' and she would reply: 'Glasgow, and proud of it!')

At the Commodore's request, Mrs Menzies also ran a laundry for the Navy at the manse – proof, she felt, that no detail was too humble for him if it concerned the welfare of his men. Or anyone else's welfare, come to that:

'One day the laird's wife said that her little girl had dropped her teddy-bear into the bay and she wished the Commodore could send his barge to rescue it as it floated about. "Woman," he shouted, "don't you know there is a war on? I've got more to do than rescue teddy-bears." – But, next day, the dripping teddy was restored to the disconsolate child, to the satisfaction of everybody concerned!'

Perhaps as a result of his traditional Naval upbringing, the Commodore was inclined to draw a sharp dividing line between men's and women's work – and he was at first very reluctant to have Wrens working under him. Knowing in advance of the Commodore's feelings, Wren Dorothy Gordon was horrified to find she'd been drafted in 1942 to HMS *Western Isles* – the headquarters ship itself, anchored in the bay, and that she was to be the only Wren on board:

'I must admit I was scared of the Admiral at first, I was so green! However, he was very kind and patient, although he had a rather disconcerting habit of coming out on deck and shouting: "Damsel, damsel", instead of ringing when he wished to give me some work.'

Dorothy Gordon was still working on board the *Western Isles* in June 1943, when an announcement was made which delighted all at Tobermory. Commodore Stephenson 'Puggy', 'Monkey', 'Monkey Brand' – had been knighted in the Birthday Honours, and everyone rejoiced that his work, so vital for the conduct of the war at sea, had been thus recognised.

There was a large tea-party in celebration on board the headquarters ship, at which Sir Gilbert assured those present that it was the Tobermory team as a whole and not its leader which was honoured. Dorothy was one of the Wrens who arranged the flowers for this occasion – flowers which had come from far and wide, for there was no florist in Tobermory. Just before the guests arrived, Sir Gilbert came in to inspect, and of course the piercing glance *had* to fall on one vase, whose flowers were drooping sadly. With a shock, the girls realised why – they'd forgotten to put any water in the vase. 'Get rid of them,' said Sir Gilbert, 'I'd rather have no flowers at all than those!'

The new knight was still the old familiar 'Terror'.

# 17

# Problem Boys

By the time Sir Gilbert got his knighthood in 1943, the Tobermory routine had been pretty well perfected. Ship after ship left the bay, with the Commodore's lessons well and truly learnt, to face the more deadly ordeal in the oceans. But some visitors created special problems – and these problems the Commodore approached with special relish.

Of all the ships that came to Tobermory, there was only one with which, in the Commodore's view, nothing could be done. She was a foreigner; not that this had any special connection with her failure, for many French, Dutch, Norwegian, Greek and Polish vessels graduated with flying colours from Tobermory.

The Commodore went on board this particular ship and said he was glad to see her and was quite ready to start work. 'Oh,' said the Captain, 'I think we'd like two or three days' rest before we start training.'

'Don't worry about that,' Sir Gilbert said, 'you've come here to *be* trained. There's nothing for you to do!'

He started by telling them to change billet – to move from one buoy to another. They took nearly the whole day to do it, and when they got to the new buoy they secured with hemp rope instead of a mooring wire.

No, thought the Commodore, this isn't the place for you! So he sent the ship away.

Among the foreign visitors, the Greeks were quite efficient, but a little difficult, because they just wouldn't stop talking.

One day a *Western Isles* Leading Seaman was taking a party of them in close-order drill ashore, and complained afterwards to the Commodore that they chattered all the time. Next day Sir Gilbert went along with him and said: 'Listen to me, men! All put your tongues out and keep them out.' They did as he said. After a pause, he continued: 'Now damn well talk, if you can – and carry on with your drill!'

Another time Sir Gilbert was on the bridge of a Greek ship, doing exercises in harbour. There were three officers with him – and they were all giving orders at the same time.

The Commodore said to the Captain: 'Don't you think it would be better if only you gave orders to the helm and the engines?'

'Well, sir,' he said, 'that would be a little difficult. But I'll tell you what I'll do – I'll shout louder than all the others!'

The Norwegians were a great success, and the Dutch did well – several Dutch submarines took part in the anti-submarine training programme.

One of these submarines came in one day after a refit with a new camouflage. She made a signal to the Commodore: 'What do you think of my new camouflage?' and Sir Gilbert made the ingenious reply: 'Where are you?'

The French came in considerable numbers to Tobermory, and probably occasioned the Commodore fewer linguistic problems than his other foreign 'guests'. He had been passable in French since childhood, though his use of the language was inclined to be idiosyncratic in the Churchillian fashion:

'I felt great sympathy with the French, who were fighting under the command of foreigners and did not know whether there would be a France at the end of the war; nor did most of them know whether their families and friends were alive or dead. All their simple possessions were in their ships, which might well be sunk one day. I asked the French Admiral to allow each officer and man a chest or cupboard at French headquarters where they might keep their most treasured possessions, but it was not granted. Their lives seemed to revolve around two focal points – the cook and their wines. Not far from Tobermory there were beds of mussels at the outlet of a small stream; the French found them, and must have eaten many thousands. Their system of punishment was simple – either they got their allowance of wine, or they did not. I accused one Captain, who was a great friend, of having transformed one of his magazines into a wine vat, but he denied the charge!'

Sir Gilbert's French victims carried his reputation back to Paris, and in 1948 his work for the French Navy was acknowledged with the Rosette of the Legion of Honour, bestowed on Sir Gilbert by the President of the French Republic.

One French Naval officer, F.D. de Fonbrune, who had visited Tobermory in the corvette *L'Aconit* (*The Aconite*) summed up the feelings of many in the French Fleet:

'I know of no better friend of France than you, sir. Your reputation is nearly as great in the French Navy as it is in the British Navy – that is saying something. Your many French friends will rejoice, as I do now, when they hear that at last you have been rewarded for all you have done for us.'

De Fonbrune wrote a book on his wartime experiences called *Combats sur Mer*. He recalls how *L'Aconit* went to Tobermory, after commissioning, to 'learn her job of defence, hunt and attack'. Though the accent is French, the story is familiar:

'A flag officer, Vice-Admiral Sir Gilbert Stephenson (retired) has been in charge of this base since 1940. It is he who, after your training, declares you "fit for combat". Or he may label you "unfit" and never hesitates to do so, for this demon of a man, as sharp as a needle, judges you rapidly and rarely makes a mistake. The extreme conscientiousness with which he takes his responsibilities forbids him to send into action any ship that has not every chance of winning. His eyes, sometimes twinkling, sometimes serious, soon discover your weak points and he sets to work to correct them.

'He tests the officers with awkward questions. "There are parachutists coming down on that island over there. What are you going to do?" And, to a bewildered Midshipman: "The magazines are on fire. Jump to it!" "You, First Lieutenant, two large landing craft are out of control. You must take them both in tow – how will you do it?" Or again: "Your derricks are broken and you must lift a ton weight on to the forecastle – how do you do this?"

'At the end of fifteen days, during which all the equipment

has been tried out and a large number of possible events envisaged, the Admiral makes a final inspection. He talks to the crew in picturesque French and wishes the *Aconit* good luck. Then I leave the little bay with dignity, saluting him as I go – confident of my ability and hoping with all my heart to meet the enemy.'

De Fonbrune's wishes were granted. Not long after leaving Tobermory on one of his two visits, he met two enemy U-Boats in one night and sank them both.

If all his students had been as apt as de Fonbrune, the Commodore would never have had to embark on the practice of removing from their ships officers he thought unfit for positions of command at sea under the demanding circumstances of the convoy war. Such a practice, without Admiralty authority, was indeed unconventional, and caused considerable surprise to newcomers to *Western Isles*.

One young Midshipman went into the wardroom on his first day in the ship to discover a mixed bunch of officers, more than one of whom was, to say the least, drowning his sorrows. He then learnt that most of those present had been the First Lieutenants or even the Commanding Officers of ships working up at Tobermory: 'They had, however, been found wanting in some respects by someone called "Monkey" and forthwith removed by him for further training. I was full of my divisional course training and could not understand how the Admiralty could allow such unilateral action – without even a court martial! I was told that "Monkey" could do anything – and frequently did; furthermore, he didn't bother to report his actions, thereby saving a lot of tiresome inquiries!'

This seems to be a fair reflection of the Commodore's attitude, and more than once the Admiralty told him he was exceeding his powers in removing officers. They also reacted strongly to the adverse reports that were sometimes sent through from Tobermory on an officer's performance. It seemed the conventional thing was to put 'Very Good Indeed' (VGI) as the Commanding Officer's comment on a report. The Commodore did not agree:

'What was the use of putting VGI on every report? I told the Admiralty that until I had a letter from the Second Sea

Lord telling me not to tell the truth, I proposed to continue to tell it!

'In many cases I had to take officers out of ships, not because there was anything wrong with them but just because they were worn out – they were exhausted.

'One Commanding Officer came to me, looking very tired, and after I'd been talking to him a bit I said: "Do you sleep quite well?"

' "No, I don't," he replied.

' "Do you dream a good bit?"

' "Yes, I do, sir, I find it rather difficult to sleep."

'I looked at him and said: "Now I'm going to tell you something rather sad. I'm going to take you out of your ship and you're going on leave for three months."

'He absolutely broke down – and he wasn't the only one. That, of course, told me that I was more than right. Many of these fellows were well overdue for a break. It was fair neither to them nor to the ship or their ship's company that they should remain in a job which they were unfitted to do, not through any fault of their own – in fact they were exhausted from doing their jobs *too well!*'

After a time the Admiralty decided to let the Commodore pursue his own methods at Tobermory, unusual though they were; and, to cope with any changes that became necessary, the Admiralty maintained a pool of officers in *Western Isles*, some under training, some awaiting appointments. Here's an example of how the system worked:

One day a frigate came in and the Captain was talking to the Commodore about his officers. When he reached the Sub-Lieutenant, he said:

'He's sick at sea and useless in harbour.'

'No use to you at all?'

'No use,' said the Captain.

'All right; send him over to me and I'll give you another Sub-Lieutenant.'

Next day the 'useless' one reported to the Commodore, who said: 'Well, lad – you didn't get on very well in your ship. What was the matter? When did the trouble start?'

'Oh, sir, it was nothing to do with the ship. It was me. I'm a fool. I know I'm a fool, my family know I'm a fool, my schoolmasters always told me I was a fool – you can't do anything about it, sir!'

'Well,' Sir Gilbert said, 'I'm not interested in what other people said about you. Now, which part of your initial training did you dislike least?'

'Well, I was quite interested in squad drill on the parade ground.'

'In that case we'll give you a little of that to do here. Go down below and put on gaiters, and in a quarter of an hour I'll send you ashore with a Petty Officer – he'll be in charge!'

Sir Gilbert had a word first with the Petty Officer, and told him not to tax the young man too much, but to keep an eye on him.

When the Petty Officer returned for dinner, he told the Commodore he thought the Sub had the makings of a good officer in him.

'All right,' said Sir Gilbert. 'I'll give him to you for the rest of the week. Make what you can of him.'

About three months later, the Sub wrote to tell Sir Gilbert that he was happily settled in his new ship and was enjoying life very much indeed and getting on splendidly: 'He was perfectly all right, that young man. He had acted like a fool simply because people told him he was a fool!'

The lessons learnt in HMS *Western Isles* tended to stick. Lieutenant Geoffrey Budd, RNVR, was one who wrote in appreciation:

'I have been instructing in *Nimrod*, striving to bring out personalities as well as imparting the necessary knowledge. Particularly satisfying is the officer or rating who scrapes through a course in such a desperate state of nerves that one wonders if he will ever hold down a job, and who later comes back as a sound member of a ship's company. In each of these people I see myself, floundering uncertainly in WI.'

Another *Western Isles* flounderer who found his feet was Sub-Lieut. Duncan Carse, RNVR, later to achieve fame as a writer, an Antarctic explorer and also as an actor on radio and television.

Carse was sent to *Western Isles* with a miserable Naval record behind him. Even the Commodore was baffled by his apparent inability to carry out the simplest duties on board. However, 'Monkey' took the trouble to find out where the Sub's real talents and interests lay, and without hesitation arranged for him to be drafted to more congenial work.

Some time later came a letter from a Duncan Carse who had found himself:

'After spending a year scripting for the Admiralty Film Unit, I was demobbed and set up on my own as a freelance film and radio man . . . I suggested it would be a good idea to make a full-length documentary on square-rig. The result is that I am now standing by for the four-mast barque *Passat* . . . we are taking timber out to the Cape and should be about eighty days on passage. With any luck we shall get some good stuff.

'You will appreciate how much it means to me to be on the move again, and how lucky I count myself to be able, in this age of mechanised decadence, to get back to a more genuine life which I know and like.

'Once again, thank you for all you did for me at Tobermory – no "difficult" Naval officer could have had a more fortunate vetting.'

During his period in HMS *Western Isles* Duncan Carse used to produce radio plays on the ship's closed circuit radio system. On one occasion they had prepared a special programme, and the Commodore was to open it with a message to the ship's company. It was due to start at 11:00 and all was ready long before. Then at 10:55 a valve failed. By 11:02 they were on the air again, but it was not good enough. The Commodore turned and said, without wrath: 'I shall not speak. If I can't be on time, how can I expect them to be?' and that was that.

During the five-year period of Tobermory's wartime activity, there were one or two serious accidents. A Chief Petty Officer had both hands blown off when a faulty fuse caused a $1\frac{1}{2}$ pound block of TNT to explode prematurely, and Petty Officer Motor Mechanic Harry Polach was killed: he is buried in a Tobermory churchyard.

Wartime conditions made many aspects of the work risky. The Commodore was terrified there might be an accident with the ship's boats at night, because they had to get about without lights; the buoys weren't lit and, of course, neither were the ships: 'How we didn't have a boat sunk was an absolute marvel – I knew we might lose a hundred men in a night if a big liberty boat had a collision and went down'.

With these hazards in mind, Sir Gilbert was concerned about his own barge, and asked the First Lieutenant to choose the crew carefully: 'In *Western Isles* we had either men old enough to be grandfathers, or young boys; and the old men were so nervous that this in itself constituted a great danger, so I told the First Lieutenant to give me a crew of boys – the sort who would have a fast motor-cycle – because their dash was far less likely to be dangerous!'

If there were few serious accidents, there were a fair number of near-misses.

One Sunday morning two depth charges were accidentally fired from their throwers in the course of drill, into the harbour, where they exploded, with a great deal of noise. The Commodore was in church at the time, and at the end of the service, took the parade as if nothing had happened, before tearing across the harbour in his barge to find out what had happened. He asked for a report on damage to the ship and to the sister ship moored alongside. The Captain of the ship concerned said he had tried to find out how the cartridges had been left in the throwers, when the Bridge tell-tale showed them as safe, but the Commodore stopped him because there was bound to be a Court of Inquiry.

The next question was typical of 'Puggy' – 'What did you do then?' to which the Captain replied: 'Carried on with drills, sir!' That was the right answer. And there was no Court of Inquiry. In the Commodore's words: 'I never said anything about those things – they were accidents; why report them to the Admiralty?'

Another explosive story created diplomatic problems.

One day a mine was reported floating down the outside of a small island, so 'Monkey' sent the torpedo gunner of the Submarine Depot ship to deal with it. He blew it up too near the island, and caused a good bit of damage there and in the town itself.

As most of the land on the island was owned by the laird, the Commodore went to make his peace, that same afternoon.

When he arrived there was a tea-party on and Sir Gilbert was invited to take a seat.

At a suitable moment he said to the laird, Colonel Allan: 'I think, Colonel, there was a little atmospheric disturbance, wasn't there, this morning?'

'Atmospheric disturbance!' the Colonel exclaimed. 'I don't know how much of my property has been damaged – and I suppose you'd call it an Act of God!'

'You've said it, Colonel!' 'Monkey' said and held out his hand. 'By God, it *was* an Act of God!'

Not long after this, a large bill arrived from the town clerk for damage done. The Commodore wrote to him:

'Thank you very much for the compliment in addressing this letter to me, but it should be addressed to the Almighty and, in spite of what you think, I am not the Almighty; the laird has told me in front of witnesses that this was an Act of God!'

No more was heard about the matter, and who paid the bill remains a mystery.

There was yet a third 'bombardment' which fortunately damaged no one.

Some Commandos training ashore wanted to stage a night attack on the Tobermory Home Guard, who agreed to take part in the exercise. 'Can't we join in, too?' asked 'Monkey'. 'We could have a vessel or two out to try and catch your boats coming over.' The answer was 'No' so the Commodore took no more notice – it was their show.

On the evening in question several loud bangs came from the direction of the town, but 'Monkey' knew what was happening and, as it was nothing to do with him, took no further notice. But then he heard a bang in the harbour where the ships were, and this was to do with him! He immediately called for his barge and off they went. The first ship they came to was a trawler, with rifles pointing over the side, ready to repel boarders. They had earlier spotted a Commando boat in the harbour, and the Captain had ordered it to come alongside at once, or be blown out of the water. The Commandos had obeyed – and the Captain, having identified them, let them go – but meanwhile they had managed to fix a small practice mine to the ship, which had, of course, gone off.

The Commodore had to deal with another considerable diplomatic problem when he found himself in disagreement with the First Lord of the Admiralty, A.V. Alexander, who visited Tobermory twice. The First Lord was very keen to get more escort vessels to sea, even if it meant cutting down by a further 5 per cent on expertly trained crews. The Commodore invited a number of Captains on board to dine with him, and after dinner he put the question to them:

'The First Lord of the Admiralty has a proposal to make to you, and he would like to know the answer. Suppose you were given charge of a big convoy and you had the choice of having twelve escort vessels, thoroughly well trained and known to each other, or twenty vessels, not known to each other and not so thoroughly trained. Now, remember, the First Lord has other things to think of than our problems. He's the man who will have to stand up and answer questions in the House: "First Lord, how many escorts did that convoy have that was so badly mauled the other day?" He would much rather answer: "Twenty" than "Twelve". But if you were in charge of the convoy, which would *you* choose – twelve fully manned and trained, or twenty not quite "full complement", not thoroughly trained?'

'Twelve, sir!' they all shouted in chorus.

'There you are, First Lord,' said Sir Gilbert, 'there's your answer!'

A.V. Alexander and he got on very well, though the Commodore didn't agree with the First Lord's socialist views. 'Once after lunch I took him into my cabin and said: "Now, First Lord, you're to sit on that sofa and put your feet up and have a rest. There'll be no interruptions or telephone calls. As far as the outside world is concerned, you're dead!" '

Sir Gilbert himself very nearly died once at Tobermory. An American merchant ship had gone ashore on the rocks outside the harbour. It was very bad weather, so the Commodore rushed out in one of the MLs to see what he could do. The plan was that the motor launch would come very briefly alongside the ship – she couldn't stay, it was far too rough – and the Commodore would jump and catch the ladder: 'I caught the ladder but I was in heavy sea-boots and I couldn't climb up. There I was, just hanging from the end of the ladder – the crew of the merchant ship did nothing to help me.

'But I had complete confidence in the Captain of our ML, which came round again – close in to the ship – and someone reached out and gave me a shove up so that I could catch hold of the rail and haul myself up topsides. If he hadn't done that, I should have been drowned for a certainty.'

Happily, the Terror was preserved to complete his work. Nothing would have tempted him to abandon Tobermory till the war was done, least of all an offer from the USA. The American Admiral who was in charge in European waters in fact invited him to go to America to train their escort ships: 'But I told him I was under orders from the Admiralty; I would go where *they* told me to.' (So long as it suited Sir Gilbert, of course.)

# 18

# Time Off at Tobermory

The first thing to say about time off at Tobermory is that there wasn't very much of it for anyone, least of all the Commodore. Only once did he take a real break away from *Western Isles* – in the summer of 1942 – and then he was as reluctant to go as were some of the Captains and First Lieutenants he considered too exhausted to continue in their ships. Admiral Sir Percy Noble, Commander-in-Chief Western Approaches, approved warmly of his plan to take a period of leave: 'That you thoroughly deserve a rest', he wrote, 'is the opinion of all who know your work in that most important Tobermory.' So south the Commodore went for a little while, to top up those apparently inexhaustible batteries.

If long periods of leave were otherwise ruled out, he did make a point of going ashore each afternoon: 'You see, I was a sort of sultan in Tobermory; no one could equal my authority – they even thought I had the Admiralty totally under my control. (I didn't bother to deny it!) But the danger of getting a swelled head was decidedly there, so I would often go up to Aros House to play with the Allans' children. I grew very fond of them and they of me, I think, and they couldn't have cared less who I was!'

Playing with the children was not the Commodore's only occupation at Aros House. Another was well described by a shipmate:

'Have you ever seen a terrier digging out a rat hole? The

expression is tense, the body quivers and the limbs work furiously, making the earth and dust fly. Walking one day in Aros grounds, I was amazed to see the air round a clump of rhododendrons full of flying leaves and branches – as though it had been struck by a tornado. Pieces of wood flew up to heaven and whirled there like autumn leaves in a gale. In the heart of the cyclone, wielding an axe, could be seen a figure with arms and legs moving so fast they seemed to vibrate in a haze.

'It was the Commodore.'

If all the visiting ships had known of the Commodore's close friendship with the Allans of Aros House, it's doubtful if the following incident, recalled by former Sub-Lieutenant RNVR Geoffrey Noble, would have taken place:

'One day a signal was received in HMS *Vervain* addressed to all ships in harbour, stating that an "Aros Sitting Hen" – apparently a valuable bird – had been stolen, and it was evident from the tone of the signal – "Commanding Officers will report forthwith, etc. etc." – that the Commodore was not amused by the incident. No one appeared to know anything about the crime, but that evening two of the Commodore's Staff Officers dined on board *Vervain*. Now roast chicken was not part of the standard menu of a North Atlantic corvette, and, to the horror of the wardroom, the steward proudly bore in a dish of roast chicken – obviously the proceeds of a successful landing party the night before. Our two guests tucked into the chicken without batting an eyelid – but next day that part of the shore was put out of bounds to HM ships and remained so until the base closed down!'

With his interest in agricultural pursuits and passionate belief in physical exercise, the Commodore started an allotment, in order to encourage his 'old' men to do the same. Most of them, however, were reluctant to follow this admirable example, though some Tobermory people did give up bits of their gardens to the few who chose to relax from the strains of war in this way.

When he was not digging or uprooting rhododendrons or skylarking with the Allan children, the Commodore, clad in 'civvies', could be seen going for vigorous walks.

One day young Anne Macfarlane was dawdling up one of the braes reading a comic, when a gruff voice behind her said:

'Going to take me in tow, hey?' and she felt the Commodore's walking stick being hooked into her trenchcoat belt. Anyone watching would have seen the diverting spectacle of a highly flattered ten-year-old towing the Commodore at a vast rate of knots up the back brae!

However, more often that vigorous gentleman proceeded under his own steam, taking his daily exercise along a little path that runs round Tobermory Bay to the lighthouse.

One day he met on the front street in Tobermory two sailors in uniform. He himself was in 'civvies' and began to talk to them. In the course of conversation, he asked them what ship they were in. Not realising who he was, one of them turned and said it was 'none of his ruddy business'; immediately the Commodore congratulated the man on his good sense in not giving away Naval information!

However, another day, when he was in uniform, he met two men who, rather than take the trouble to salute him, turned and looked in a shop window. He immediately challenged them: 'I know why you did that. This is a butcher's shop and has nothing in it to interest you. You wanted to avoid saluting me.'

If, in the course of his walks, he saw a ship doing anything he thought not quite right, he would simply roar a command across the bay – no megaphone was necessary in his case. But his roaring voice made him seem more domineering than he really was; and the kind-hearted man beneath the gruff exterior would always emerge freely in the company of the young.

All who were children in Tobermory at that time remember the wonderful Christmas parties he put on for them. No one knew where the marvellous crackers, toys and cakes were found, but the Navy would always come up with a party to end all parties, in spite of wartime austerity. The Commodore did not delegate attendance at these parties to junior officers but was to be seen there 'Grand Old Duke of Yorking' with the best.

Christmas and the New Year was a good time at Tobermory. Ships in harbour on Christmas Day actually had the day off (the only one in the year) – except that they were all expected to take part in vigorous boat-pulling races round the harbour after the festive meal.

The Commodore made a point of putting in a personal

appearance at the WVS canteen. He appeared one morning at 9 o'clock just before the New Year and after a few questions and complimentary remarks about the services rendered by the canteen, asked one of the helpers, Mrs M.C. Macfarlane, if she had her barrel of whisky ready for the Hogmanay festivities. She replied that she never touched the stuff and wasn't likely to have a bottle, let alone a barrel – to which reply he gave a cackle of laughter and took his departure.

The Christmas spirit even extended to the 'victims' in harbour. One Captain, Lieut.-Cdr. R.C. Freaker, RNVR, when he came to Tobermory in command of a frigate over Christmas in 1942, very much wished to have his wife up there for this period. He thought the Commodore would frown on anything which might divert attention from 'working-up" – however, approval was given and Freaker's wife duly arrived at the Western Isles Hotel.

Freaker knew of the Commodore's practice of inviting the COs of working-up ships to dine with him in the *Western Isles* on about two occasions on each visit; but, on the visit when his wife was there, he was not invited to any of these 'stag' dinner parties. Instead, he and his wife were invited to lunch with him one Sunday, when his ship was down for a general drill that afternoon. When, after lunch, Freaker expressed the view that it was time he returned to his ship for the general drill, Sir Gilbert said: 'You take your wife for a walk ashore; I will look after your ship this afternoon – that's an order'. It was, of course, obeyed.

On the night before Freaker's final inspection, his ship was down for a night firing practice. It looked, therefore, as though he would be unable to spend this last night ashore with his wife. However, as he was entering harbour late that evening on completion of the 'shoot', he received a signal from the *Western Isles* saying: 'Barge is being sent now to take CO inshore' and before the ship was secured to the buoy the Commodore's own barge was waiting alongside for him: a gesture Freaker never forgot.

The Commodore was not so generous to himself in these matters, devoted though he was to his family. His wife came up about twice a year and stayed about a fortnight, but the

Commodore never slept ashore. Lady Stephenson would come on board *Western Isles* in the morning and stay in the ship for the day, then the Commodore would take her ashore in the evening and leave her at the jetty. He always made a point of taking her at the time when liberty men were coming off – to show he had confidence in them.

Tobermory, in fact, had the reputation of being the best-behaved Naval base in Scotland. Women, it was said, could walk out in the evenings and never be accosted, and there was hardly any drunkenness. But this did not mean that there was a lack of fun and games for the staff of *Western Isles* and the visiting ships' companies. Some of the most popular parties took place at the 'Wrennery' ashore, and one of them gave rise to a misunderstanding which might have endangered the high moral reputation of the base. For non-Naval readers, it should perhaps be explained that the phrase 'to take off' can mean simply 'to take someone, or something, out to a ship from shore'.

An RNVR Lieutenant came into the Commodore's cabin one day and asked if he would stop 'the appalling stories that are being told about one of the Wrens'.

'What stories?'

'Well, one of the seamen on board has a pair of Wrens' pants.'

'Who told you?'

'The engineer.'

'Ask the engineer to speak to me.'

The engineer arrived a few minutes later.

'Now, Chief, what's all this about a pair of Wrens' pants being on board?'

'Oh, yes, sir. Chief E.R.A. Barrable said a sailor had a pair of Wrens' pants.'

'Tell Barrable I want him.'

Barrable duly appeared.

'Barrable, what do you know about Wrens' pants?'

'Well, sir, at Admiralty pier at about 9.30 last night a sailor took off a pair of Wrens' pants.'

'Who told you?'

'It's the talk of the mess, sir.'

'Barrable, you realise you could be tried by court martial for spreading untrue stories about HM Forces! I want to know who told you and, if you don't remember, find out!'

'Very good, sir.'

'Monkey's' next step was to send for the Second Officer WRNS.

'Now, did you have a party in the Wrens' quarters last night?'

'We did, sir.'

'Were you there?'

'No, sir.'

'Was the Third Officer?'

'No, sir.'

'Who was in charge?'

'A Petty Officer Wren, sir.'

'Very well. You will have everybody who was at that party on board my ship at three o'clock this afternoon.'

At this point the RNVR Lieutenant who had started it all reappeared, still under the impression that an enforced striptease had taken place on Admiralty Pier.

'I hear there's to be a full enquiry, sir,' he began, with the utmost solemnity. 'The most dreadful things may come out!'

'Oh yes,' Sir Gilbert said, 'I shall endeavour to throw light on every single circumstance.'

At exactly 15:00 hours, crowds of sailors and Wrens were reported present to the Commodore – so many that they appeared to stretch well below the horizon! He questioned the Petty Officer Wren first.

'Now, Petty Officer, what exactly happened at that party last night?'

'Everything was all right, sir. Nothing untoward occurred.'

The next victim was a male Petty Officer, whose reply was similar: 'Nothing unusual, sir.'

'*Nothing*, Petty Officer?'

'Well, sir, I did see a sailor chasing a young Wren upstairs.'

'Oh, and what did you do about it?'

'Went upstairs and fetched them down, sir, and then placed a guard on the stairs.'

'Very sensible.'

And so the investigation went on. One after another the people who had been at the party were asked what had happened. Everyone denied that anything particularly scandalous had occurred. A good deal of time had passed before it was the turn of a certain Able Seaman to be questioned.

'Now look here, do you know anything about Wrens' underclothes?'

'Well, as a matter of fact, I do, sir. You see, as I was leaving the house I noticed some clothing hanging out in the garden to dry and when I went closer I noticed a pair of Wrens' pants and, just for a lark, I removed them and took them off with me from the jetty.'

'Just for a lark! Do you realise that all these people have been dragged here merely on account of your peculiar sense of humour? You will return the pants immediately – I am not really accusing you of any immoral act but please, in future, guide your sense of humour in other directions!'

Yes, however relentless the working routine at Tobermory – and it was at least a twelve-hour-a-day, 364-day-a-year business – the times of relaxation were most enjoyable. The *Western Isles* staff put on lively revues at the village hall and, both in these and at the numerous parties in ships and ashore, the Commodore was naturally the object of much of the humour. We might end this chapter with one of the numerous irreverent songs composed in brief off-duty hours by his weary victims, and sung with immense gusto in the little frigates and corvettes which would soon be engaged in a life and death struggle on the high seas:

SONGS CLEAN, OFFICERS FOR THE SINGING OF
'TOBERMORY MERRY-GO-ROUND'
(With apologies to Widdicombe Fair)

1. In vair Tobermory us worked up our ship,
    Up along, down along, panic and vlap,
    And every durned day some danged vule made a blip,
        Riggin' derricks,
        Oistin' whalers,
        Layin' kedges,
        Streamin' Voxers,
        Firin' rockets,
        Floatin' rafts
    AND MAKIN' A BALLS OF IT ALL,
    AND MAKIN' A BALLS OF IT ALL.

2. One the day when us ztreams our way in vrom the Zound,
   Up along, down along, panic and vlap,
   Us zees zum corvettes what was muckin' around,
      Riggin' derricks,
      Oistin' whalers, etc.

3. The Base Staff comes out vor to give us a blast,
   Up along, down along, panic and vlap,
   There was vifty aboard us afore us made vast,
      Riggin' derricks,
      Oistin' whalers, etc.

4. The virst one aboard was t'ole Commodore,
   Up along, down along, panic and vlap,
   If looks could've killed you'd 'ave zeen us no more,
      Riggin' derricks,
      Oistin' whalers, etc.

5. Us tried zum vield trainin' but oh what a vrost,
   Up along, down along, panic and vlap.
   The Commodore pranced and 'e shouts all is lost,
      Riggin' derricks,
      Oistin' whalers, etc.

6. In 'arbour us tried out zum general drills,
   Up along, down along, panic and vlap,
   And gave the instructor zum 'air raisin' thrills,
      Riggin' derricks,
      Oistin' whalers, etc.

7. Vor Asdics us went to the old Western Isles,
   Up along, down along, panic and vlap,
   The language us 'eard scorched the country vor miles,
      Riggin' derricks,
      Oistin' whalers, etc.

8. Us went vor a shoot and our 'Guns' was the mug,
   Up along, down along, panic and vlap,
   But back there zum other poor duffers will be
      Riggin' derricks,
      Oistin' whalers,
      Layin' kedges,
      Streamin' Voxers,
      Firin' rockets,
      Floatin' rafts,
   AND MAKIN' A BALLS OF IT ALL
   AND MAKIN' A BALLS OF IT ALL.

# 19

# Victory

Thus, day after long day, the work went on – and the little ships left the somewhat stormy shelter of Tobermory Bay one after another to play their part in the struggle against the U-boats; a struggle which, by 1944, was going very much the Allies' way – thanks in no small measure to the training procedures which had been evolved at *Western Isles*.

Among the general run of days, 20 October 1944 was a landmark, though there was no interruption in the normal routine. Just after ten o'clock that night the following signal was addressed to the Commander-in-Chief, Western Approaches, at Liverpool – and repeated to the Director of Anti-Submarine Warfare at the Admiralty:

> HMS *Clover*, the thousandth vessel to be worked up at Tobermory sailed at 22:00 today.

To this simple statement, which summed up four years of immense activity, the C-in-C, Admiral Sir Max Horton, replied:

> Congratulations on magnificent achievement. Helen of Troy's reputation was achieved with much less effort.

By way of acknowledgment the Commodore rounded off the exchange as follows:

> Very many thanks for your very kind signal which we all appreciate very much. Your reference to 'Helen of Troy' – may I say that you are the first of my friends to appreciate my face value!'

With Allied troops advancing in Europe, and the worst of the war at sea over, it became clearer every day that final victory could not long be delayed: and the Commodore was determined to make his Victory celebrations, when they came, a worthy public finale to the Tobermory achievement – something that everyone ashore or afloat would never forget.

'I began planning for VE Day six months before it came, because I didn't want any trouble. We didn't know how many ships might be in – maybe ten, maybe thirty – all full of men with their steam right up, bursting – with all the strain gone. They were going to do *something*, and if it wasn't a legal something it would be an illegal something, so I intended to have the safety valves well working for three days.

'They wouldn't have time to realise what was happening – it would all be great fun, and they would be far too exhausted to create trouble!'

The Commodore discussed his plans as they began to mature. He explained to Cdr. P.S. Jones, RN, how ships' companies sent ashore for field training would collect firewood for an enormous bonfire which would appear as a beacon over the islands. Everyone on Mull would be his guest ashore that day. Jones felt that the Victory celebrations – and particularly the beacon – were, to the Commodore, symbolical of everything he had worked for so magnificently.

The bonfire was built, and a day and night guard was placed on it as the day approached; it exceeded even 'Monkey's' expectations in size and splendour. Only one thing went wrong. A high spirited sailor fired a Very cartridge into the base of the massive wooden structure, and, since the lower foundations were primed with inflammable oil, the result was inevitable – some time before the Commodore himself was to have pressed the firing button to seal and complete his personal contribution to the war.

Whatever disappointment the Commodore may have felt at this contretemps, it was a very small matter in the context of a prodigiously successful celebration – the result of months of meticulous planning by the *Western Isles* staff. In fact, it's said that his officers were so exhausted by the organisation required that they begged the Commodore to start the war again!

There were three full days of jollification in Tobermory Bay, on 8, 9 and 10 May 1945. It all began at 8 a.m. on VE Day with sirens, ships' bells, bangers, thunderflashes, and rockets, with twenty ships of all types in the harbour – the noise was tremendous. Then they all Dressed Ship. At 10.30 a.m. there was the thanksgiving service ashore in the open air; very good by all accounts but, unfortunately, it was pouring in torrents; in spite of which the Commodore took the salute at a march past of nearly 1,000 officers and men.

After a restful afternoon, ships 'spliced the main-brace'. Everyone remained on board for the King's speech and then, at 11.15 p.m., there was a great procession of 200 torch-bearers, headed by a squeegee band, through the town and up to the top of the golf course. Here the tremendous bonfire which consisted of nearly 40 tons of wood was ablaze. On top of the bonfire was a gibbet with an 18-foot effigy of Hitler hanging on it. The fire was visible for upwards of 50 miles, and it burned for two days and nights.

The next morning, Wednesday, at 9 a.m., the sports and regatta began. First, a series of six rowing races for different classes of ratings, with an officers' race, followed by an all-comers' race with Carley floats, rafts or anything else they cared to use.

In the afternoon, there was a 10-mile walking race, a treasure hunt and a 3-mile cross-country race. In the late afternoon and early evening, two performances of an ENSA entertainment, followed by dancing on the waterfront, which was lit up with fairy lights. It seemed everyone in the Isle of Mull was there to join in – and it was all a huge success. At 11.45 p.m. the ships gave a marvellous pyrotechnic display for nearly an hour; rockets, snowflakes, cresset flares, coloured lights – with thunderflashes to represent gun-firing.

On the Thursday morning at 9 a.m. it was off again.

Obstacle races for skiffs, Wrens' pulling races, a Carley float mêlée – the harbour was alive with small boats and cheering.

In the afternoon there was the great athletic sports meeting in the recreation ground – all laid out like an Olympic track. Over 2,000 people crammed the little ground and free teas were supplied to everyone. Over 300 fine prizes were presented, and

HMS *Western Isles* also presented a magnificent silver cup to the ship aggregating the highest points – a frigate, HMS *Widemouth Bay*.

There was once again dancing on the waterfront that evening, and another magnificent display of pyrotechnics for nearly an hour – even better than the previous night – followed by a 'night-action fight' between the eleventh-century Spanish caravel *Santa Maria* and the privateer *Black Hawk*. This was really well done, with a commentary, broadcast on the waterfront. The 'ships' were old motor drifters cleverly disguised, and the effect was splendid; it was quite dark, and the action was shown by flares dropped to give a silhouette effect; as the ships, manoeuvred by wires, approached each other, thunderflashes, went off by the score, portraying gun flashes and explosions; eventually both 'ships' caught fire and burned merrily all night until they sank.

During the three days' celebration, all food on shore was free, and there was not one defaulter from any ship, although everyone was full of gaiety and good humour.

VE Day was the cue for another exchange of signals between Commodore *Western Isles*, and the Commander-in-Chief, Western Approaches. In reply to the Commodore's victory congratulations, the following signal was received from the C-in-C:

> Your kind signal is very warmly appreciated. No one realises better than I how much we owe to you and your staff for the high standards set in the initial training of escort vessels. Western Isles methods are known and admired throughout the navies of the united nations. That phrase from an ancient ditty might well be taken as your motto – quote – of course you can never be like us, but be as like us as you are able to be – unquote– my best wishes for good luck in the future to all of you.

At the core of the VE celebration were the solemnities of remembrance and thanksgiving. The Commodore character-istically chose the Future as the theme of his Victory address, after a reminder of the dark days of 1940 'when all the world except ourselves believed we had no possible hope'.

'How deeply thankful to the Almighty we should be', the Commodore went on, 'for the gift of that faith, that unity, that fellow-feeling in trouble and danger, that were so conspicuous in our people and especially so in those places and times when dangers and distress were most marked. It is to me impossible to believe that such great qualities and especially such unity could have been called forth from all, had we been fighting a war for material gain, for dominance over others, or for lust and hate. It was the righteousness of our cause which was our strength, and who can doubt but that it was our faith and our singleness of purpose which brought us safely through these long years of struggle to this great victory.

'In looking ahead and in deciding what road we shall travel, we are deciding on what shall be the fruit of our victory. Shall it be bitter, or shall it be glorious and worthy of the countless lives given to win it?

'We have seen the utter destruction of the people who set up a false God and who despised the Christian virtues of kindness, justice and tolerance. We have ahead of us difficulties and dangers as great as any that we have surmounted. May I ask you to remember the faith by which we have been enabled to carry on to Victory; to remember the comradeship of service and how it has helped us. It is clear that faith in ourselves and trust in our leaders will be needed in the future even more than it has been in the past.

'May we hope that our allies may succeed in building up the Kingdom of God on Earth and, in filling our hearts with faith in God, may cast out all fear.

'Finally let us, when choosing our road, determine by our conduct and example to help to build our new world, worthy of the sacrifice of the countless lives, worthy of the trust our God has placed in us, his people.'

The job of HMS *Western Isles* was virtually over – and, just a few weeks after VE Day, it was time for the Commodore himself to depart. It was a moment charged with emotion; one member of the *Western Isles* staff declared he was glad to be out of the ship that day for he would have hated to see the Commodore go. A day or two before his departure, the Commodore summed up the Tobermory achievement when he bade farewell to the ship's company over the ship's radio:

'I am leaving *Western Isles* next Friday and am taking this opportunity to say goodbye to you all. Personally I should prefer to do this direct, face to face, but, on the other hand, from your point of view it is better, for you are, I hope, comfortably seated in your messes.

'The work of *Western Isles* is such that the results, good, bad or indifferent, did show up. You know that the results were good. We have received as many as 560 signals of thanks from ships who are grateful for the help given them here. Besides that, the First Lord of the Admiralty and our Commander-in-Chief, Admiral Sir Max Horton, have commended your work and I feel that we can say with truth that our ship has won a great name. This name has been gradually and slowly built up by the enthusiasm and consistently good work of each one of our instructors and of our very expert maintenance staff. They all made it a point of honour that, as each ship sailed, those on board her knew their work, and her equipment was in order. We always had our instructors on board when practices were going on so that mistakes could be pointed out at once: it was in the consistent, expert and conscientious work of our instructors that *Western Isles* differed from other bases.

'Our work, of course, has been to try and get our ships ready to fight, or to be able to carry out any work entrusted to them. That means that not only must the equipment be in the pink of condition but that everyone must know their own work and must know it in the dark as well as in the daylight. The manner in which the gunnery has been worked up almost from scratch till a ship could take on four targets at a time and hit most of them – that all the communication ratings have been taught and practised in their work – that the radar – that most delicate and important equipment – has been made to work and the operators taught how to use it – and the way machinery defects have been found out and made good, has been due to the skilful and conscientious work of the whole personnel of *Western Isles*. And, perhaps not least, I should mention our close-order drills which have helped so many to become accurate and instant in obedience. How Tobermory will miss the kind voice of Mr Mitchell in their main street teaching his men the gentler arts of war! Speaking about the gentler arts brings me to the Wrens. How well they have upheld the Service at Tobermory – and I think you will agree with me that the visual team have given many men an urge to increase speed of engines.

'Our successful work here has depended upon many factors; among them stand out the boats. The coxswains in particular have done fine work – the Commander-in-Chief has personally commended four of them.

'Another most important job has been the staff office work. The plans of the ships' programmes, when we had ten to twenty ships, was a mystery to me and one mistake would put it all out. We did *not* have mistakes – and very great credit is due to those responsible. Another very important factor has been the friendly and good behaviour of our ship's company. Their example has been a very valuable asset. I don't think that many bases have as good a record as Tobermory, nor have they as happy memories as we can carry away with us. Keep it up and help Commander Talbot as you have helped me to build up and maintain to the end the good name of *Western Isles*.

'I would like to name my many excellent friends, starting with Commander Cann, Commander Hugonin and Commander Talbot, who, I am glad to say, is to continue with the good work of *Western Isles*. Right from the first day to this, I have had the utmost support in every way from Lieutenant-Commander Palmer. His skill, resource and great experience have always been forthcoming. It is difficult to picture what I should have done without him and his care in organising the ship's entertainments, and his thought generally for all about him has helped to render the work here as little burdensome as it well could be.

'In a few days I am leaving the Service for the second time, a Service in which I have spent most of my life; and if I had to spend twelve more lives in this world I would choose that each one was spent in the Navy. When I first came to sea as a Naval cadet it was rubbed into me by the Senior Midshipman that, in future, the Service was to come first and any family or private affairs to take second place, if any. If I may presume to give you advice it is that you follow that line; do your job to the utmost of your strength, and you will find great happiness.

'To those of you whom I may not see again, I say – Thank you – and goodbye.'

Soon, HMS *Western Isles* ceased to exist save in the memory of those who had served in her or had been her 'victims' in the long years of war. Now the struggle was over and all were free to depart, to pursue happiness in their own individual ways. And yet surely more than one Tobermory man would have echoed the Commodore's brisk reply when, not long after the VE celebrations, he met a friend in the confused bustle of Glasgow's Argyll Street at rush hour:

'What on earth brings you here?' asked the friend. Answer: 'It's this *blasted* peace!'

# Epilogue: Retirement
## (1945–1972)

# 20

# Stand Up, Sit Down

So this fiery episode in the little town of Tobermory had come to an end and it was left to subside gently into its accustomed Hebridean peace. The public saw some of the charms of the Isle of Mull in the film *I Know Where I'm Going*, which had its beginnings towards the end of the Commodore's reign, and once the war was over, people in general began to learn something of what had been happening over the previous five years in Tobermory Bay.

By the time the base closed down, 1,132 'work-ups' had been given to 911 different ships (some returned more than once). Tobermory graduates were known to have destroyed 91 U-boats and probably sank 38 more, and they disposed of 39 enemy aircraft plus possibly three others. 'Some 200,000 men', in the words of one newspaper account, 'passed through this base to enter the sternest battle the Navy has ever had to fight . . . few men have exercised more influence on the war in the Atlantic than Commodore Stephenson. He knows small ships and has U-boat warfare at his fingertips. But, above all, he has one supremely valuable gift – he transmits his enthusiasm to every officer and man with whom he comes in contact.'

Sir Gilbert's work, as we have said, had been acknowledged with the knighthood bestowed on him in June 1943, though at that time security forbade any detailed citation – and his name was renowned in the Royal Navy. But now it was all over. Was it, perhaps, time to retire quietly, at the age of 68, from public pursuits? Or would a new focus be forthcoming for those formidable talents?

Well before Tobermory closed down, the answer was in the making; the Commodore (about, once again, to revert to his true dignity of Vice-Admiral, Retired) was already being pressed to become a member of the newly formed Sea Cadet Council, and not long after quitting HMS *Western Isles* in 1946 he was injecting his unabated enthusiasm into this young organisation for young people. For three years Sir Gilbert served on the council, to such effect that, when Admiral Sir Lionel Halsey died, he was invited to assume the office of Honorary Commodore of the Sea Cadet Corps, a title which he held for nine years from November 1949 to November 1958, when, a mere eighty years of age, *he* thought (though everyone disagreed) that he was no longer quite equal to the job as he saw it.

The trouble was, of course, that to Sir Gilbert the word 'honorary' did not carry the usual connotation of a somewhat sedentary dignity. He spent his nine years in the job rushing up and down the country visiting each and every sea cadet unit which invited him. In fact, he never refused such an invitation. Everywhere he went he brought inspiration and encouragement, though his methods were, as ever, sometimes unexpected.

His inspections were always searching, and he made a point of speaking individually to every boy. After all this, he usually made a speech to all those present.

Sir Gilbert followed this procedure on a visit to the Isle of Man Unit in 1956. The boys were duly seated in the hall of Castle Rushen High School, Castledown, and the CO of the unit escorted him in. The boys stood up. The Admiral stepped on to the platform and told them to sit down. No sooner had they sat down when he shouted 'Stand up!' In fact he did this about twelve times in as many seconds. When they finally sat down Sir Gilbert said: 'You will now be sufficiently awake to listen to me.'

This technique was frequently employed by the Admiral, and not only to gain the attention of boys. On one occasion he combined a sea cadet inspection with giving the prizes at Speech Day at Bedstone School in Shropshire. It was a fine day and the parents were seated on the lawns – the usual collection of large summer hats and a number of clerical collars. When the moment came for Sir Gilbert to address the parents there was a long pause until suddenly, at the top of his voice, Sir Gilbert shouted: 'Stand up! Sit down! Stand up! Sit down!' The astonished parents eventually obeyed, and there was the remarkable sight of the dignified audience jumping up and down – much to the delight, naturally, of the boys. After three or four such movements Sir Gilbert shouted: 'Now I hope you are all awake, especially you clerics, and will listen to what I have to say!'

Similar tactics were employed when the Admiral visited the Tooting unit of the sea scouts in South London, of which he was President for many years. On one occasion he chose to drop in unannounced, wearing civilian clothes. The sight that met his eyes did not please him; the officers and cadets happened to be just standing around, apparently doing nothing. At once he gave orders for the officers to remove their caps and jackets, and then conducted a ten-minute PT session, demonstrating the exercises himself!

Each year the Admiral went to Tooting to present the unit trophies. A former officer of the Unit, Lieut. P.J. Mite, RNR, remembers the routine:

'The Commanding Officer told us that Admiral Stephenson always arrived bang on time, even if it meant driving round the block for a few minutes beforehand. True enough, at exactly 7.30 a small man, smothered in gold braid and medals, was piped aboard. He inspected the entire unit and spoke to every cadet on parade. He then watched a short display of various activities before "Stand Easy".

'After "Stand Easy" he presented the cups to the various cadets and had a quick word with each one. At the end of the presentation he addressed the unit as a whole.

'Admiral Stephenson's address was always informative and he never failed to pay tribute to the officers and instructors of the unit whose work was all too often taken for granted. He did not give us a talk on the Royal Navy and its traditions and discipline, but on our part in life and our duties to our families and the community. He always stressed the need for leaders in present-day life and encouraged us to make the most of sea cadet training, which could develop these qualities.

'He always finished his address by telling a short story, always beginning: "I know it's all right to tell you boys this because my vicar told it to me."

'On the two occasions that Sir Gilbert visited the unit after I had become an instructor, I was greatly impressed by his manner in the wardroom. He was not the self-opinionated person that boys might expect a person of such high rank to be, but was a very modest man – far more interested in us than himself. His conversations were not about his experiences and work during his Service life, but about such things as the way barmen wash glasses in pubs compared with the way he does the washing up in his own home.'

Sporting activity was, as Sir Gilbert saw it, a vital part of the sea cadet's life and here, again, the Hon. Commodore performed his duties to the full. One day, early in the 1950s, he presided at a sea cadet football final between a London unit and a Liverpool unit, held on the Nottingham Forest ground.

'We had arranged', writes Lieut.-Cdr. Frank McKay, RNR, 'for him to meet the Nottingham Forest committee before the match and to "inspect" the two teams, the referee and two linesmen, on the field and the match was due to kick off at 3 p.m. We had allowed five minutes for this. Before the match, too, we had arranged for the Peterborough unit band to provide the music before the match started and at the interval. Sir Gilbert shook hands with the football players and officials, and the referee duly placed the ball on the centre circle, with both teams ready and waiting to do battle immediately Sir Gilbert reached the touchline. The Forest ground is rather a wide one, and the Peterborough band was still assembled at the touchline near the entrance to the directors' box. The band was, of course, brought to attention and full honour given to Sir Gilbert. He did no more than decide to inspect the band. Never in the history of bands was a band more thoroughly inspected! Every cadet bandsman had a "personal chat" and an inspection of his instrument. The game had obviously to be held up and was started about 25 minutes late. All was forgiven afterwards.'

While Sir Gilbert was Commodore, the National Boxing

Championships of the Sea Cadet Corps grew annually in importance, with the finals held under Royal patronage at the Royal Albert Hall. The presence of Sir Gilbert was a great encouragement to all, and sometimes drew to the hall people not addicted to pugilism. It is said that a former Tobermory man once applied to the secretary, who was somewhat surprised, for a ringside seat. 'But, John,' he said, 'I didn't know you were interested in boxing!' 'I'm not,' came the reply, 'it bores me stiff, but I want to see once more the Seaman of the Century!'

For Sir Gilbert, one of the greatest objects of the corps was to instil into boys 'that smartness, self-discipline and spirit of service which have characterised British seamen from time immemorial', and in this way to make him 'the "proper man" – sailor, citizen and good fellow'. Discipline was essential, for 'you cannot bring any human activity to success, if it involves more than one person, *without* discipline'.

But at the core of the Admiral's concept of the corps was a profoundly religious view of life. The Secretary of the Preston Sea Cadet Unit has not forgotten the moment, after a sailors' service in the town, when Sir Gilbert pulled him into a pew to discuss the unit's progress, with the words: 'Where better to talk together of things close to our hearts than in the House of God?'

When Sir Gilbert decided it was time for him to leave the sea cadets, his departure was greatly regretted on all sides and it brought a letter from Prince Philip, who was well acquainted with all he had done for the young lads of the country:

Dear Admiral,

I am very sorry to hear of your retirement and I am sure the whole Corps will miss your enthusiasm and wise leadership.

No one will ever be able to tell you how valuable your services have been to the Sea Cadets, but I hope that in the years ahead you will have plenty of time to reflect on a highly successful turn of office.

Yours sincerely,
Philip.

Plenty of time! Sir Gilbert's eighties and nineties were as full of pugnacious activity as any other period of his life. As the 'officer through whose hands there passed more RNVR officers than any other', he always had a very special place in the hearts of Naval reservists and was unsparing in his visits to Reserve units. The Croydon Unit of the RNVSR (Supplementary Reserve) was one of a number to benefit from his patronage; in fact, in this case he was a founder member: 'I thought it would be a grand idea', says Lieut. John Broughton, RNVR, 'to start off the new unit with a gorgeous supper or a dinner at a local hostelry. The remark was overhead by "Monkey", who wheeled round: "Dinner? A new unit? Am I invited? Can I come?" From that moment I realised I had a fourth member and the most important support of all.

'He loved young people; boys, sea cadets, sons of the lads who had served with him at Tobermory . . . and he had an immense regard for and faith in the young of today. I remember him insisting that I bring my son to lunch with him at the United Services Club one day. Christopher was then a rather dishevelled and inky fourteen, but was suitably awed and impressed by the decor and ambience of the important club. A rather difficult meal was got through with 'Monkey' punctuating each mouthful with a beautiful question shot at Christopher.

'These prodding quizzes can be very alarming to the uninitiated; my wife and I attended a dinner of the Western Approaches Reunion at the Dorchester. When it came to our turn "Monkey" looked pityingly at Joyce and murmured: "You poor child – did you know him before you married him?"

'Coming up behind us was a cleric with his campaign miniatures. The padre stepped forward with the usual "Good evening, sir". "Monkey" looked at him: "What are you doing now?" – to which the Reverend replied: "Well, sir, I am Rural Dean of Nottingham," (or some such town). "Oh, and what are you doing about the Sea Cadets?" – broadside from "Monkey", digging the Padre in the turn. "Sea Cadets?" he replied, "Well, sir, to tell you the truth . . ." Tell me the truth? Don't you always speak the truth?" roared "Monkey".'

Lunch with Sir Gilbert at the 'Senior' (the United Services Club) was quite an experience. Apart from enjoying the excellence of the food and drink, his guests were often left gasping, as Commander R.W.D. Thomson was, by the Admiral's curious sense of humour:

'On the way in to the dining room, he went up to another member, poked him in the chest and, in a loud voice, asked him: "Still got the same wife?" The man looked too startled to be able to reply, whereupon the Admiral turned to all within earshot and, again in a loud voice, said: "There you are, you see – doesn't even know!" And then repeated, emphasising and separating each word – *"doesn't – even know!"* We then passed on, leaving the poor man looking utterly bemused. In the dining room I asked the Admiral who he was. He grinned, and replied: "No idea – never seen him before in my life."'

And there's something very characteristic about this happening in the City recalled by Commander R.H. Bristowe, RN, who had led an escort group at Tobermory:

'Five years later, the Admiral had not forgotten us. The Stock Exchange firm which I had joined had several sailors from my ships, and he asked us to buy some Government stocks. For this type of investment the contract note is sent out "for cash" instead of "for settlement on account day". I was away from the office two days later when the Admiral arrived with a carpet bag. The office manager politely tried to relieve him of it, but the Terror of Tobermory was having none of it. It contained £1,000 in one pound notes from a cheque he had just cashed to comply with the contract note! He wanted to count it out.'

Much less wel known than quirky tales of that kind are the numerous examples of kindness known to the Admiral's many friends, but naturally not publicised. And his ability simply to cheer people up has not in the least diminished with age. As one ex-RNVR officer wrote to him in 1966: 'I was feeling a bit despondent when I arrived in London – a feeling of having failed coupled with the feeling that success was going to elude me. But all that evaporated as soon as I saw you. Thank you for everything! What a pity you didn't go into the Church; a blunt humour might have stopped you from becoming Archbishop of Canterbury, but you might have got your own diocese and then, I'm sure, the Church would have had some kick in it!'

Short of being ordained, the Admiral did in fact contribute a great deal to the Church. He was very active in the concerns of the parish church at Saffron Walden and was a notable lay preacher. He frequently talked to children about religion and was a great success with them, as he was with adult gatherings, in church or elsewhere.

Some of the Admiral's ethical beliefs were unorthodox; for example, he advocated euthanasia and was sympathetic to Spiritualism, and on these and other topics he would argue with anyone – a bishop was perfect prey!

There is no doubt that the years at Tobermory were, for the Admiral, the high point of his career on this earth. He kept in touch with many who served under him there, and each autumn held a tea-party at the 'Senior' where many of them gathered to talk of old times – and new. Each year, too, he made a point of attending the Battle of the Atlantic Service in Liverpool Cathedral, where he read the Naval prayer in a voice still rich and steady at four score years and ten or more; one doesn't often hear the Bible read with such clear authority. His presence on these occasions and so many others was, for those who remembered, a sharp reminder of the work on which so much depended in the Bay of Tobermory and in the wider waters of the Northern oceans.

In his mid-nineties, the Admiral still darted about with remarkable agility, living the good life as he saw it and enlivened the world in the process. In the autumn of 1970 he wrote a brief Declaration of Faith, which sums up his beliefs – beliefs that would be difficult now for all to share. But of course they were formed when Britannia ruled the waves – and much else besides:

'The Christian, God-loving atmosphere I absorbed in my childhood, with pride in our country and its great men, and the general love and admiration for Queen Victoria by all Englishmen, has remained with me all my life; there was also in me the strong feeling that courage, straight dealing and hard work were the natural gifts of all Englishmen, and that it was the function of the English to rule.

'These beliefs have been a part of me all my life.

'And then, when I joined the Royal Navy, I became conscious that, of all Services in the world, the Royal Navy stood first and that it was the very bulwark and defence of our great country. I think that, later on, the story of Clive and his Indian soldiers at the siege of Arcot sealed and confirmed my views. You may recollect that, in about 1748, when the British and the French had rival business interests in India and our establishments were on the verge of capture by the French, Clive – a clerk 25 years old – organised our forces and, with a body of 120 English soldiers and 200 Sepoys, seized the town of Arcot which was besieged by a French force of 10,000. During this long siege his native soldiers came to Clive, who was almost at his wits' end, not to complain of their short rations but to offer rice – their only food – to the English, content to keep for themselves the water in which it had been boiled. What greater proof of devotion could be found in all history?

'My unchangeable view is, and always will be, that the English have done more good to the world, not only by their deeds but also by their example, than any other race, and that in all the countries they have governed they were respected and loved by 90 per cent of the population, especially in India, and that the only unkind act of ours was to leave them.

'May I say that my opinion of the courage and high character of the English people was amply confirmed by my five and a half years' experience at Tobermory.'

# PostScript

So how does one end this story of a man who lived a happy life, surrounded by friends and inspired by the same beliefs which inspired him from earliest youth?

Sir Gilbert Stephenson gave expression to those beliefs when he was invited to open HMS *Eagle*, the new headquarters of the Royal Naval Reserve in Liverpool, in the summer of 1971: he was then 93 years of age.

'It is possible,' said the Admiral on that occasion, 'that there are several persons present here today who have experienced the feeling that their country was in very great danger, not by reason of the loss of a battle on land, not by reason of the loss of a battle in the air, but because our sea communications were threatened. If an enemy wishes to conquer us, not a man need be landed, not a bomb dropped on our territory: cut our sea communications, and we are left with a choice of starvation or surrender.

'Some of you may remember what that great leader Lord Jellicoe said: "I am the only man who can lose the war in an afternoon" – and Sir Winston Churchill once declared: "The only time I have known fear was when the Battle of the Atlantic was at grave risk."

'Surely it is up to those who love their country and appreciate its situation to do all they can to persuade young men and women to join the Royal Navy or its Reserves. I can give you an additional reason for this: if they join the Navy they are on the right road to happiness.

'May I illustrate this by a little incident at Tobermory. When I was saying farewell to one of our escort vessels, I said, in addressing the assembled ship's company: "Are you happy?" Back came the roared reply: "Yes, sir, we are!" I then continued: "Now I'll tell you why you're happy. You are all, each one of you, of one mind. You intend to help your ship to play its part in winning the war. You know your work – you mean to do it with all your might and main – and you know that if you fall out, your friends to right and left of you will carry on your job. You have nothing to hide from anyone, you can look everybody full in the face.

' "Now I'm going to ask you to imagine something. Suppose that your Captain has called you together and has said to you: 'I have great news – the war is over! Peace is declared!' Suddenly your whole line of thought is changed. You look down, so that no one can see what you are thinking: you will be wondering whether Jack over there will get the job you want.

' "For God's sake, when peace does arrive, remember your happy time together in war: and continue to think of your country and others as well as yourself." '

Those words, spoken after twenty-five years of peace, in the port which had once been the headquarters of Western Approaches Command, express the essential Gilbert Stephenson. A man whose principles are summed up in an inscription which hung over his bed, written in the French he learned as a boy at his mother's knee:

*'La vie n'est pas un plaisir ni une douleur, mais une affaire grave dont nous sommes chargés, et qu'il faut conduire et terminer à notre honneur.'*

'Life is neither a joy nor a sorrow, but a serious responsibility laid upon us, which must be fulfilled and brought to an honourable conclusion.'

# Index

## Index